ADVANCES IN TRAFFIC P

Human Factors in Road and Rail Transport

Series Editors

Dr Lisa Dorn
*Director of the Driving Research Group, Department of Human Factors,
Cranfield University*

Dr Gerald Matthews
Professor of Psychology at the University of Cincinnati

Dr Ian Glendon
*Associate Professor of Psychology at Griffith University, Queensland,
and President of the Division of Traffic and Transportation Psychology
of the International Association of Applied Psychology*

Today's society confronts major land transport problems. Human and financial costs of road vehicle crashes and rail incidents are increasing, with road vehicle crashes predicted to become the third largest cause of death and injury globally by 2020. Several social trends pose threats to safety, including increasing vehicle ownership and traffic congestion, advancing technological complexity at the human-vehicle interface, population ageing in the developed world, and ever greater numbers of younger vehicle drivers in the developing world.

Ashgate's Human Factors in Road and Rail Transport series makes a timely contribution to these issues by focusing on human and organisational aspects of road and rail safety. The series responds to increasing demands for safe, efficient, economical and environmentally-friendly land-based transport. It does this by reporting on state-of-the-art science that may be applied to reduce vehicle collisions and improve vehicle usability as well as enhancing driver wellbeing and satisfaction. It achieves this by disseminating new theoretical and empirical research generated by specialists in the behavioural and allied disciplines, including traffic and transportation psychology, human factors and ergonomics.

The series addresses such topics as driver behaviour and training, in-vehicle technology, driver health and driver assessment. Specially commissioned works from internationally recognised experts provide authoritative accounts of leading approaches to real-world problems in this important field.

Advances in Traffic Psychology

EDITED BY

MARK SULLMAN AND LISA DORN
Cranfield University, UK

CRC Press
Taylor & Francis Group
Boca Raton London New York

CRC Press is an imprint of the
Taylor & Francis Group, an **informa** business

CRC Press
Taylor & Francis Group
6000 Broken Sound Parkway NW, Suite 300
Boca Raton, FL 33487-2742

First issued in paperback 2017

© 2012 by Mark Sullman and Lisa Dorn
CRC Press is an imprint of Taylor & Francis Group, an Informa business

No claim to original U.S. Government works

Version Date: 20160226

ISBN 13: 978-1-4094-5004-7 (hbk)
ISBN 13: 978-1-138-07297-8 (pbk)

Visit the Taylor & Francis Web site at
http://www.taylorandfrancis.com

and the CRC Press Web site at
http://www.crcpress.com

Contents

List of Figures

List of Tables

PART I
Driver Personality, Emotions and Stress

Chapter 1

Driven by Anger:
The Causes and Consequences of Anger
during Virtual Journeys

Amanda N. Stephens
The University of Edinburgh, UK

John A. Groeger
University College Cork, Ireland

Introduction

Anger is a unique emotion because in many ways it contrasts with the influence of other negative emotions. Fearful or sad drivers may be more likely to drive cautiously and adopt safer driving behaviours (for example, Arnett et al., 1997). Angry drivers, however, can become aggressive and tend to incorporate a higher level of risk-taking into their driving style (Deffenbacher et al., 2001). This may be because anger encourages confidence and when angry, we can overestimate, or fail to estimate, the amount of control we have in specific situations and make quick, stereotypical judgements (Lerner and Keltner, 2001). The dangerous effect of anger is worsened in the driving context because anger-driven aggression and risk-taking have very real consequences, both for the driver, any passengers in the vehicle and for other road users. Angry drivers have been found to be twice as likely to be involved in traffic accidents during simulated driving scenarios (Deffenbacher et al., 2003). In real traffic conditions, self-reported anger has been related to near miss accidents (Underwood et al., 1999) and excessive speed choices (Arnett et al., 1997). For these reasons, it is important to understand what makes drivers angry, how this anger is expressed and who the potential targets of driver aggression may be.

In this chapter we discuss the contributory roles of trait and state factors in driver anger and subsequent aggression. We present an overview of research we conducted while both affiliated with the University of Surrey (UK), using a STISIM 400W fixed-base full-car interactive driving simulator. We address questions relating to the situation specificity of trait influences and examine how affect-related judgement biases influence evaluations of, and behaviour in, various driving situations. We conclude with a discussion of the implications of our research for traffic safety and future research needs.

What Makes Drivers Angry?

Individual Differences in Trait Anger

Much of the research on driving anger tends to focus on anger as an inherent disposition. For example, researchers often use the Driving Anger Scale (DAS; Deffenbacher et al., 1994) to measure driving-anger 'traits' and compare these tendencies against self-reported behaviour or other individual characteristics. The original DAS contains 33 items which assess anger proneness across six types of traffic situations: events involving slow, discourteous, or illegal driving by other motorists; traffic obstructions; the presence of police vehicles and; receiving hostile gestures from other drivers.

There is some uncertainty as to the most appropriate factor structure of the DAS. Researchers have argued for a five-factor model which combines items for slow driving and discourteous driving into one factor (N = 166; O'Brien et al., 2002). Sullman (2006) proposed a four-factor model for his data on New Zealand drivers (N = 861), suggesting trait anger specific to situations of risky driving by others, progress impediment, discourtesy and hostile gestures. Lajunen et al. (1998) found data from a sample of British drivers (N = 280) fitted a three-factor model of anger propensities across situations of reckless driving, impeded progress and direct hostility. Björklund (2008) supported this three-factor model in a sample of Swedish drivers (N = 98). Elsewhere, we have avoided the uncertainty regarding DAS factor structure by relying on total DAS scores as a measurement of trait driving anger (Stephens and Groeger, 2009).

The variability in the number and structure of the factors underlying the DAS leads to ambiguity over what situations provoke driver anger. A 'trait' implies a consistent manner of appraisal and or behaviour, rather than something which necessarily extends across all situations. Thus, it is important to agree on what situations provoke anger and then understand who is more likely to become angered. When individual differences in DAS scores are found, it is usually that females report more anger across various driving situations (Björklund, 2008; Deffenbacher et al., 1994). However, gender differences seldom emerge on overall DAS scores (Deffenbacher et al., 1994; Stephens and Groeger, 2009). Overall trait driving anger appears to decrease with age (Sullman et al., 2007) and be unrelated to how often one drives (Deffenbacher et al., 2003).

We investigated the factor structure of the DAS using a sample of British drivers (N = 192; mean age = 22.82 in years; SD = 5.53; males = 96). Our drivers were affiliated with the University of Surrey and participated as a prerequisite for selection in further research. Our aim was to understand what situations are conducive to trait driving anger and which individual characteristics relate to these traits. To test this, we first performed a confirmatory factor analysis (CFA) using the robust maximum likelihood procedure in EQS (v. 6.1 for windows; Bentler, 2004) on scores for the 33-item Driving Anger Scale. We then examined the relationships between the resultant factors and age, gender, motivational orientation (measured with the General Causality Orientation Scale, GCOS; Deci and Ryan, 1985) and perceptions of driving skill (measured with the Self-Assessed Driving Skill Questionnaire; Groeger and Grande, 1996) as well as state anger (Profile of Mood States-Shortened Bilingual Version; Cheung and Lam, 2005). Finally, we assessed associations between trait anger and behaviour during a 20-minute simulated drive designed to measure judgements of speed and distance (based on work by Groeger et al., 1999).

The CFA showed support for the original six-factor model of driving anger. All goodness of fit indices were acceptable: Santorra-Bentler χ^2 (471) = 665.06, Comparative fit index (CFI) = .90, Root Mean-Square Error of Approximation (RMSEA) = .04, 90 per cent confidence interval (CI) of RMSEA = 0.04: 0.05. Table 1.1 lists the means for each driving anger factor. There were no gender differences across any of the driving anger situations, nor on overall trait driving anger (t < 1 for all except traffic obstructions where t = -1.01, p > .05). Inter-correlations, adjusted for multiple comparisons, and conducted on driving anger traits, age and driving experience (years licensed and miles driven per week and per year) also revealed no individual differences in anger propensities. The exception was that older drivers and those licensed for a longer time were more likely to report anger over illegal driving by others (r(192) = .25, p < .001 and r(192) = .24, p < .001, respectively). Since age and licence length are themselves correlated, partial correlations were conducted and revealed no reliable relationships between age and illegal driving and licence length and illegal driving when the other was controlled for.

Propensities to become angered across the six situations were also largely unrelated to motivational orientation and perceptions of skill (own and that of a novice driver). However, two exceptions emerged. Drivers scoring lower on autonomous orientation, and therefore less likely to take responsibility for their own situations, reported more probable anger over hostile gestures (r(192) = -.19, p < .01). Drivers considering themselves to be less skilled reported greater tendencies to be angered by police presence (r(192) = -.17, p < .05).

Table 1.1   Self-reported driving anger across six situations by gender

Situations of:	Males (N = 96) M (SD)	Females (N = 96) M (SD)	T (190)
Traffic Obstructions	2.26 (.57)	2.34 (.72)	<1
Illegal Driving	2.47 (.91)	2.58 (.90)	<1
Slow driving	2.42 (.56)	2.49 (.67)	<1
Hostile Gestures	2.98 (.93)	2.99 (.93)	<1
Discourtesy	3.19 (.59)	3.23 (.63)	<1
Police Presence	1.84 (.82)	1.74 (.70)	<1
Total DAS	2.52 (.46)	2.56 (.51)	<1

We also compared trait anger to state anger measured after the drive. We found moderate relationships between driving anger traits and angry mood (r(192) = .28, p < .001), although these results do little more than highlight a relationship between angry mood and the likelihood of reporting anger across various situations. When considered across the six driving anger situations, anger over police presence (r(192) = .24, p < .001), discourtesy (r(192) = .28, p < .001) and to a lesser extent, slow driving (r(192) = .19, p < .01), illegal driving (r(192) = .18, p < .01) and traffic obstructions (r(192) = .16, p < .05) were positively related to current reported mood. Mood was not significantly related to anger over hostile gestures (r(192) = .09, p = . 26).

How is Anger Expressed?

Self-reported anger propensities are associated with increased reports of road-way aggression and a larger number of actual traffic accidents (Deffenbacher et al., 2000; Deffenbacher et al., 2003). To assess whether trait or state anger relates to tendencies to speed or run red lights we asked drivers to make a number of speed, distance and time to collision (a combination of speed and distance) judgements while driving in an unadorned driving environment (see Stephens, 2008). We measured how often drivers exceeded posted speed limits and examined event-related behaviours, which included number of occasions when drivers drove through an intersection displaying a red traffic light signal as well as variation of driving speeds while following lead vehicles travelling at either 30 or 50 miles per hour (mph). We also investigated unassisted speed production judgements of 30 mph and 50 mph. Correlations of the driving variables showed no strong relationships. The two speed production variables shared the largest correlation (r(192) = .51, p < .001). Speed production in slow speeds was positively related to variation while following slower drivers (r(192) = .16, p < .05); while it was unrelated to variation while following vehicles travelling at 50 mph. Speed production of 50 mph was related to less variation when following at 50 mph. Drivers more likely to exceed the speed limit were also more likely to run red lights (r(192) = .30, p < .001).

Stepwise regression analyses were conducted to assess whether age, the six driving anger subscales or state anger predicted driver behaviour. It is important to note that some of the DAS subscales have strong correlations with each other (discourtesy and traffic obstructions r = .58; discourtesy and slow driving r = .65) and will, therefore, mask each other's relationships with driving behaviours. State and trait anger factors failed to feature in any of the most robust models (see Table 1.2), with the exception that anger over slow driving accounted for a small percentage (< 5 per cent) of the variance in speed variation during enforced following of a lead vehicle travelling at slow speeds. Anger over illegal driving also accounted for less than five per cent of the variance in the production of slower speeds (30 mph) suggesting that drivers more angered by illegal driving drive slower in lower speed zones.

Table 1.2   **Age, Trait driving anger and State Anger as predictors of driving behaviours**

Predictor Variables	R square	Adjusted R square	R square change	P	Standardised β weight	Correlation with behaviour	Partial Correlation
Speed compliance							
Age	.04	.04	.04	<.001	-.19	-.19	-.19
Variation of following speed 30 mph							
DAS Slow Driving	.03	.02	.03	<.05	-.16	-.15	-.16
Variation of following speed 50 mph							
Age	.03	.03	.03	<.05	.16	.16	.16
Speed Production 30 mph							
DAS Illegal	.03	.03	.03	<.05	-.17	-.17	-.17

The lack of reliable relationships between anger-traits and general behavioural tendencies provides some tentative evidence that evaluations moderate the relationship between trait and behaviour. Anger-prone drivers are more likely to see the frustrations and annoyances of the current situation (Deffenbacher et al., 2000). However, the drive discussed above was novel in that it contained a series of trials where for the most part, no interaction with other road users or objects was possible and thus few annoyances could be [mis]interpreted by drivers. In the few situations involving other drivers, such as the follow tasks, we found an association between traits and behaviour when the lead driver was driving slower.

Are Trait and State Influences General or Situation Specific?

While state and trait influences on driver anger have been considered across a number of various driving situations, few studies relate these at a situation-specific level. We addressed this issue by comparing trait driving anger to anger evaluations and behaviours measured while driving (Stephens and Groeger, 2009). Twenty-four drivers (mean age = 25.45 years, SD = 7.47; males = 12) drove 10 miles through urban, rural and residential areas during which they encountered everyday driving situations ranging in the level of anger-provocation. These included, but were not limited to, jaywalking pedestrians, slower lead vehicles and oncoming traffic encroaching into the driver's lane. Drivers were requested to make evaluations of their current levels of anger and frustration throughout the drive. These occurred at predetermined locations, either immediately after a driving event or after a period of uninterrupted driving. Speed and lane positioning were also measured at these locations. A subsequent group of drivers (N = 24; mean age = 25.41 years; SD = 7.42; males = 12) matched on age, driving anger propensities and driving experience completed an identical drive, but were required to rate the level of danger and difficulty of each driving scenario.

The results of this study provide evidence for the specificity of driver anger. We separately correlated for each driver his/her anger evaluation and behaviour in the three-second period before and then after each evaluation was required and then averaged the individual correlations as per Dunlap et al. (1983). Anger evaluations were higher immediately after situations requiring drivers to have extreme changes in their speed ($r(430) = -.13$, $p < .05$). Trait driving anger did not moderate this relationship. Correlated averages showed that overall trait anger scores were strongly related to general tendency to evaluate driving situations as anger-provoking ($r(24) = .58$, $p < .01$). However, when anger propensities were examined across situations of high and low provocation, drivers higher in trait driving anger tended to evaluate low anger situations as anger provoking ($r(24) = .65.$ $p < .01$), but trait anger was relatively unrelated to anger evaluations of high anger-provoking situations ($r(24) = .34$, $p < .05$). Thus, regardless of trait anger dispositions, drivers become angry when impeded. However, drivers more likely to become angry while driving do so in non-provoking situations.

We also found that the situational-specificity of trait anger was evident with driver behaviour. Although drivers higher in trait driving anger reported more driving violations (measured with the Driver Behaviour Questionnaire; Reason et al., 1990) ($r(46) = .42$, $p < .01$), DAS scores were unrelated to actual general behaviours, such as overall speed or number of collisions across the drive. In low-anger-provoking situations, trait anger mediated the relationship between anger and subsequent behaviour. Drivers higher in trait driving anger, when angrier, drove faster and moved more within their lane. These results suggest that drivers prone to anger present more of a threat in low-anger situations.

Of further interest is that comparison between the two groups of drivers (anger focus versus danger focus) revealed differences in general driving behaviours. Drivers focusing on the inherent threat within each scenario drove in a more cautious manner, with more consistent speeds and less erratic braking patterns than drivers making continual anger evaluations.

When considered alongside results of the larger study discussed earlier, our findings argue for a situation specific model of driving anger that relies on evaluations of the situation. Trait-propensities appear to provide a general cognitive bias which leads anger-prone drivers to evaluate ambiguous and non-provoking situations as anger-provoking and react with aggressive behaviours. This may be due in part to the fact that anger propensities share, albeit weak, relationships, with perceptions of driving skill and the tendency to not take responsibility for one's situation.

Our pattern of findings is consistent with cognitive-motivational-relational theories of emotion (see Berkowitz, 1990). This model suggests anger results from a two stage situation-specific appraisal process that first assesses how relevant the situation is to one's goals and then (a) whether there is a target of blame; (b) the individual can cope with the situation; and (c) the expectations of the outcome. Thus, anger trait characteristics may influence evaluations of ambiguous situations. However, in situations where a driver's journey is disrupted by an obvious perpetrator, it is the evaluation of this disruption that leads to anger and subsequent behaviour.

Is 'Who' More Important than 'Why'?

To understand what factors are important when making anger evaluations of traffic situations we conducted two studies in which we manipulated characteristics of a target of blame. We did this by altering the status and culpability of a slower lead driver and examining the effects of this on reported anger and behaviour (Stephens and Groeger, submitted).

Status assessments, determined by relative skill, control over rewards or punishment or social standing, are often used to regulate anger and subsequent aggression in other contexts (Allan and Gilbert, 2002). Higher status individuals are more likely to blame lower status group members for negative situations and to react aggressively toward them. There is some evidence to suggest status also regulates behaviour in the driving situation. Observational studies have found drivers to be more aggressive toward vehicles of lower socio-economic status after situations of unjust impediment (McGarva and Steiner, 2000).

In Study I, we contrasted reactions to slower lead vehicles that were either marked with a learner plate (low status) or unmarked (control) when the behaviour of these vehicles was within, or determined by an event obviously beyond, control of the lead vehicle. In Study II, reactions to ambulances (high status) or otherwise identically sized generic work vans (control) were compared. For each study, participants (N = 24, males = 12) drove for 20 minutes in a suburban environment where they encountered three types of following tasks: one of each with a status vehicle, and an identical set with a control vehicle. The three types of follow tasks, counterbalanced throughout the study, varied in levels of anger provocation and culpability of lead-driver behaviour. These included a low impediment (where a driver drove in front of the participant without impeding their progress), high legitimate impediment (where progress was slowed due to external risk factors) and high illegitimate impediment (where progress was slowed for no apparent reason). Subjective evaluations of driver anger (verbal anger evaluations) were taken while drivers drove, immediately and then again sometime after (precisely 2,000 feet) each follow task. These were compared with behaviour (speed and proximity to other vehicles) measured during the follow task and the subsequent 2,000 feet of uninterrupted driving.

Lead Driver Status Influences Anger Evaluations

Across both studies we found that status influenced reported anger. A repeated measures (2×3×2) ANOVA assessed the effect of status (status or control), anger provocation and culpability (low anger/no impediment, medium anger/high legitimate impediment, high anger/high illegitimate impediment) and duration of effect (in other words, immediately after following or delay after following) on driver anger. Drivers in Study I (mean age = 22.85, SD = 3.81) reported higher levels of anger after following a low status vehicle than after following a similar unmarked sedan (F(1,23) = 4.37, p < .05; see Figure 1.1). When the lead vehicle slowed or changed course because of the actions of another road user, drivers were reliably more angered when slowed by a learner driver than when slowed by an unmarked sedan (F(1,23) = 5.90, p < .05). In contrast, in Study II, drivers (mean age = 21.04, SD = 4.91) were less angry after following a high status vehicle than the similar sized work van (F(1,23) = 3.82, p = .056). Drivers were also angrier after following a slower van when there were no obvious reasons for erratic behaviour than after being impeded by an ambulance exhibiting identical illegitimate behaviours (t(23) = 3.50, p < .01). These results show that learner drivers appear to be blamed more readily for circumstances beyond their control, while higher status vehicles are more readily forgiven for unexplained behaviour.

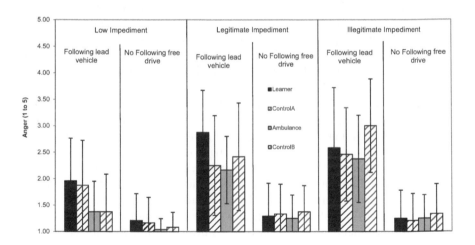

Figure 1.1   Driver anger as a function of other-driver status, nature of impediment and following/no following across Study I and Study II

Behaviour Differs According to Lead-Driver Status

Other-driver status also influenced driver behaviour. Repeated measures analysis (status × provocation and culpability) was performed on driver behaviours of minimum time to collision and minimum distance while following slower drivers. In Study I, a reliable interaction between other-driver status and provocation and culpability of impediment was found (F(2,42) = 3.39, p < .05). Paired t-tests comparing the absolute minimum time to collision drivers allowed between themselves and learner plated sedan and between themselves and the unmarked sedan showed that drivers drove dangerously closer to learner drivers (M = .40 second, SD = .15) than the unmarked vehicles (M = .47 seconds, SD = .16) in situations where there was no apparent cause for the other driver's behaviour (t(21) = 2.74, p < .05). In Study II, the absolute minimum time to collision did not differ according to status or impediment (Fs < 1). However, overall, generic work vans were followed more closely (M = 28.28 feet, SD = 11.55) than higher status vehicles (M = 30.83 feet, SD = 10.89; F(1,23) = 7.10, p < .01), indicating that drivers were more likely to drive closer to the anger provoking work-vans, whilst allowing more distance between themselves and the ambulance.

A 2×3×2 analysis (status × nature of impediment × delay) was conducted on driver speed. Drivers in Study II drove reliably faster and in excess of the 40 mph posted speed limit for some time after the most anger-provoking following tasks, where the behaviours of the lead drivers could not obviously be explained (M = 42.52 mph, SD = 7.13) than after similar impediments during justifiable circumstances (M = 39.75 mph, SD = 5.75; F(2,46) = 5.23, p < .01). This provides preliminary evidence that aggression may arise from events that have occurred sometime before the aggression is displayed.

Do Anger and Aggression Carry Over in to Subsequent Driving Situations?

Much of the research on driving-related anger and aggression has focused on the extent of anger reported while driving where the circumstances in which people drove were likely to elicit anger, or on the aggressive behaviours which result. However, anger can affect judgement in circumstances which would not themselves elicit anger. This 'incidental' anger (Lerner and Keltner, 2001) can result in misattributed blame (Quigley and Tedeschi, 1996) and the potential for enhanced aggression can also be increased by situation-irrelevant angry mood (Lerner et al., 1998).

Some driving research suggests that anger which is incidental, rather than integral, to the driving circumstances influences both mood and behaviour. Thus, for example, angry mood has been associated with worse driving performance when performance was criticized by an accompanying driving instructor (Groeger, 1997). Angry drivers display more risk taking behaviours in subsequent simulated driving, by driving closer to lead drivers, weaving more within their lane and having less steering wheel recovery on hazardous driving events (Garrity and Demick, 2001). Mood elicited while driving has also been found to affect later circumstances. Thus, for example, negative mood resulting from particular driving experiences increases the likelihood of frustration when later performing tasks unrelated to driving (White and Rotton, 1998). Thus, albeit indirect, there is some empirical evidence to suggest that the influence of mood unrelated to driving and mood elicited by driving may both extend to later driving and non-driving activities.

We conducted a study to investigate whether anger and aggression arise because of events prior to those circumstances in which the anger is experienced and the aggression is exhibited (Stephens and Groeger, 2011). Participants (N = 96, mean age = 22.44, SD = 5.41; males = 48) were allocated to one of four groups (N = 24; males = 12) statistically similar in age, driving history and propensity to become angered. All participants drove twice in the simulator. During the first drive (lasting approximately 15 minutes) participants experienced one of the four combinations of time pressure (with/without) and enforced following of a slower lead car (impeded or unimpeded) determined by group allocation. We then assessed whether affect and behaviour during a subsequent, non-provocative drive would change accordingly and whether evaluations of driving circumstances similar to, and unlike, those in which the anger was provoked would alter according to driver mood. During the second drive, lasting approximately 15 minutes, drivers followed slower lead drivers and were required to respond to events not encountered during the provocative drive. These events included jaywalking pedestrians, crossing in front of the driver and oncoming vehicles encroaching into the driver's lane. Mood (profile of mood states) was assessed before and after each drive and anger evaluations and behaviour (mean and variation of speed, lane positioning and number of collisions) were measured simultaneously while drivers drove.

Repeated Impediment and Time Pressure Lead to Increased Anger and Aggression

Anger increased and both mood and driving behaviours deteriorated in drivers exposed to slower lead vehicles and time pressure when compared to control drivers. Two-way between subjects (time pressure × impediment) ANOVAs showed that angry mood increased for drivers forced to follow slower lead vehicles ($F(1,91) = 13.09$, $p < .001$) and for drivers under time pressure ($F(1,91) = 5.07$, $p < .05$). Three-way ANOVAs on anger and behaviour were conducted with between groups factors of impediment (impeded by lead vehicle × unimpeded by lead vehicle) and time pressure (with or without) and within subjects factor of nature of follow task (no impediment, consistent impediment, inconsistent impediment). A reliable three-way interaction ($F (2,184) = 8.19$, $p < .001$) revealed that drivers reported more anger after being consistently impeded. Drivers' speed ($F(1.92, 176.91) = 30.98$, $p < .001$) and lane positioning ($F(1.75, 161.63) = 34.72$, $p < .001$) immediately after the impediment was also influenced by the nature of the preceding impediment. Drivers had more varied speed but more consistent lane positioning when this was preceded by slower lead drivers.

Anger-Congruent Behaviours Transfer into Subsequent Drives

Behavioural differences of speed and lane positioning carried over into the subsequent drive even to situations unlike those in which the original provocation had occurred. A reliable three-way interaction ($F(1,92) = 7.74$, $p < .01$) showed that drivers whose progress was slowed in the previous drive, drove with less consistent speeds while following a lead driver not providing impediment, as if they were anticipating being slowed. Immediately after enforced following, variation of speed continued to differ according to previous impediment ($F(1,92) = 4.70$, $p < .05$), but not according to recent impediment. Thus, drivers who had been repeatedly slowed during the provocation drive, had more varied speeds when unimpeded in the subsequent drive, regardless of whether a lead vehicle had recently impeded their progress.

The carry-over effects were not contained to circumstances similar to the provocation drive. Drivers who had previously been impeded by slower lead drivers not only attempted to steer around potential sources of impediment, which was evidenced by more variation of speed ($F(1,92) = 6.02$, $p < .05$) and lane positioning ($F(1,92) = 3.97$, $p < .05$) they also approached pedestrian road hazards with less caution, resulting in a greater number of traffic collisions than previously unimpeded drivers ($t(94) = -2.97$, $p < .01$). Thus, repeated provocation in one drive lead to less safe behaviours in a subsequent, non-provoking drive.

Conclusion and Implications for Traffic Safety

Across our studies we found that individual differences in anger are situation specific and, accordingly, the extent to which situational-characteristics provoke anger relies largely on how they are evaluated by individuals. Factors such as anger propensities, skill beliefs and motivational orientation may influence evaluations leading to anger in non-provoking situations. In anger provoking situations, it is not always what is happening that leads to anger, but who is causing it. Drivers become more angered and drive dangerously close to lower status drivers, when status is determined by relative skill. Angry moods resulting from driving demands also promote quick stereotypical judgements, leading to less safe behaviours, including faster driving speeds, less compliance with posted speed limits, a higher number of collisions and the tendency to follow slower drivers at dangerously close distances. These behaviours transfer into subsequent driving situations, explaining why anger and aggression may sometimes appear unprovoked and unrelated to the current driving circumstance.

If anger and subsequent aggression rely mainly on situational evaluations, rather than the situations themselves, there is scope to reduce unsafe behaviours by targeting driver appraisal tendencies. We found some evidence to indicate that changing the focus of what a driver thinks about influences their behaviour. In our study, drivers focusing on anger were more aggressive compared to those focusing on safety yet driving under the same simulated traffic conditions. Other researchers have also shown reappraisal can lead to a reduction in driving speeds (Matthews, 2002). Further research is required to understand the most efficient way of achieving this goal.

Acknowledgements

The second author wishes to acknowledge the funding support of Science Foundation Ireland (09/RFP/NES2520) and Ireland's Road Safety Authority.

References

Allan, S., and Gilbert, P. (2002). Anger and anger expression in relation to perceptions of social rank, entrapment and depressive symptoms. *Personality and Individual Differences, 32,* 551–65.

Arnett, J.J., Offer, D., and Fine, M.A. (1997). Reckless driving in adolescence: 'State' and 'Trait' factors. *Accident Analysis and Prevention, 29,* 57–63.

Bentler, P.M. (2004). *EQS for Windows, 6.1.* Encino, CA: Multivariate Software, Encino, CA.

Berkowitz, L. (1990). On the formation and regulation of anger and aggression: A cognitive-neoassociationistic analysis. *American Psychologist, 45,* 494–503.

Björklund, G.M. (2008). Driver irritation and aggressive behaviour. *Accident Analysis and Prevention, 40,* 1069–77.

Cheung, S.Y., and Lam, E.T.C. (2005). An innovative shortened bilingual version of the Profile of Mood States (POMS-SBV). *School Psychology International, 26,* 121–8.

Deci, E.I., and Ryan, R.M. (1985). The general causality orientations scale: Self-determination in personality. *Journal of Research in Personality, 19,* 109–34.

Deffenbacher, J.L., Deffenbacher, D.M., Lynch, R.S., and Richards, T.L. (2003). Anger, aggression, and risky behaviour: A comparison of high and low anger drivers. *Behaviour Research and Therapy, 41,* 701–18.

Deffenbacher, J.L., Huff, M.E., Lynch, R.S., Oetting, E.R., and Salvatore, N.F. (2000). Characteristics and treatment of high-anger drivers. *Journal of Counselling Psychology, 47,* 5–17.

Deffenbacher, J.L., Lynch, R.S., Oetting, E.R., and Yingling, D.A. (2001). Driving anger: Correlates and a test of state-trait theory. *Personality and Individual Differences, 31,* 1321–31.

Deffenbacher, J.L., Oetting, E.R., and Lynch, R.S. (1994). Development of a driving anger scale. *Psychological Reports, 74,* 83–91.

Dunlap, W.P., Jones, M.B., and Bittner, A.C. (1983). Average correlations versus correlated averages. *Bulletin of the Psychonomic Society, 21,* 213–16.

Garrity, R.D., and Demick, J. (2001). Relations among personality traits, mood states and driving behaviours. *Journal of Adult Development, 8,* 109–18.

Groeger, J.A. (1997). Mood and driving: Is there an effect of affect? In T. Rothengatter, and E. Carbonell Vaya (Eds.), *Traffic and Transport Psychology, Theory and Application.* Oxford: Pergamon.

Groeger, J.A., Carsten, O.M.J., Blana, E., and Jamson, H. (1999). Speed and distance estimation under simulated conditions. In A.G. Gale, I.D. Brown, and C.M. Taylor (Eds.), *Vision in Vehicles VII.* Amsterdam: Elsevier.

Groeger, J.A., and Grande, G.E. (1996). Self-preserving aspects of skill. *British Journal of Psychology, 87,* 61–79.

Lajunen, T., Parker, D., and Stradling, S.G. (1998). Dimensions of driver anger, aggressive and highway code violations and their mediation by safety orientation in UK drivers. *Transportation Research Part F, 1,* 107–21.

Lerner, J.S., Goldberg, J.H., and Tetlock, P.E. (1998). Sober second thought: The effects of accountability, anger and authoritarianism on attributions of responsibility. *Personality and Social Psychology Bulletin, 24,* 563–74.

Lerner, J.S., and Keltner, D. (2001). Fear, anger and risk. *Journal of Personality and Social Psychology, 81,* 146–59.

Matthews, G. (2002). Towards a transactional ergonomics for driver stress and fatigue. *Theoretical Issues in Ergonomic Science, 3*(2), 195–211.

McGarva, A.R., and Steiner, M. (2000). Provoked driver aggression and status: A field study. *Transportation Research Part F, 3,* 167–79.

O'Brien, S.R., Tay, R.S., and Watson, B.C. (2002). An exploration of Australian driving anger. In Proceedings *2002 Road Safety Research, Policing and Education Conference* 1, Australia, pp. 195–201.

Quigley, B.M., and Tedeschi, J.T. (1996). Mediating effects of blame attributions on feelings of anger. *Personality and Social Psychology Bulletin, 22,* 1289–88.

Reason, J.T., Manstead, A.S.R., Stradling, S.G., Baxter, J.S., and Campbell, K. (1990). Errors and violations on the road: A real distinction? *Ergonomics, 33,* 1315–32.

Stephens, A.N. (2008). *The role of the individual, the situation and previous driving conditions in experienced and expressed driving anger and angry mood.* (Doctoral dissertation). Available from http://epubs.surrey.ac.uk/theses/466/

Stephens, A.N., and Groeger, J.A. (2009). Situational specificity of trait influences on drivers' evaluations and driving behaviour. *Transportation Research Part F, 12,* 29–39.

Stephens, A.N., and Groeger, J.A. (2011). Anger congruent behaviour transfers across driving situations. *Cognition and Emotion, 25*(8), 1423–38.

Stephens, A.N., and Groeger, J.A. (submitted). Following slower lead drivers: Lead driver status moderates the anger provocation and subsequent exoneration.

Sullman, M.J.M. (2006). Anger amongst New Zealand drivers. *Transportation Research Part F, 9,* 173–83.

Sullman, M.J.M., Gras, M.E., Cunill, M., Planes, M., and Font-Mayolas, S. (2007). Driving anger in Spain. *Personality and Individual Differences, 42,* 701–13.

Underwood, G., Chapman, P., Wright, S., and Crundall, D. (1999). Anger while driving. *Transportation Research Part F, 2,* 55–68.

White, S.M., and Rotton, J. (1998). Type of commute, behavioral after effects, and cardiovascular activity: A field experiment. *Environment and Behavior, 30,* 763–80.

Chapter 2

Urban and Rural Differences in Attitudes Related to Risky Driving Behaviour: The Role of Sensation Seeking and Risk Perception

Matthew Coogan
The New England Transportation Institute, USA

Sonja Forward
VTI, Sweden

Jean-Pascal Assailly
INRETS, France

Thomas Adler
Resource Systems Group, USA

Introduction

Motor vehicle crashes are a significant cause of mortality in the US (where our study population is located): 34,000 Americans died in 2009 in motor vehicle crashes (NHTSA, 2010) and death from a motor vehicle crash is the single largest cause of mortality for both male and female Americans between ages 5 and 35. The death rate among young men is over twice that of women of the same age. Drivers aged 16–19 are four times more likely to crash than are older drivers (National Center for Health Statistics, 2007). Preliminary data show that residence of the driver in a fatal crash is also a key factor contributing to mortality from auto crashes. National Fatal Accident Report System (FARS) data show that the mortality rate from auto crashes for American males under 30 in the least dense (most rural) quintile of the population is four times as high as the rate in the densest (most urban) fifth of the population (NETI, 2010).

Literature and Theory

This paper presents the results of a multi-group structural equation model (SEM) which incorporates several key theories concerning the correlates of risky driving behaviour, examined to reveal differences between the urban and rural populations. Our final model explicitly integrates key elements of the Theory of Planned Behaviour with the concept of denial of risk, and the personality trait of sensation seeking. The Theory of Planned Behaviour (TPB) (Ajzen, 1991) has provided a theoretical basis for a wide variety of research into driving behaviour and other health outcomes. The TPB has been the theoretical underpinning in over 1,000 published articles on a wide variety of health outcomes (Fishbein and Ajzen, 2010), and has also been used to predict a range of different transport related behaviours (see Bamberg et al., 2007; Budd et al., 1984; Chan et al., 2010; Elliott et al., 2003, 2005; Forward, 2009, 2010; Letirand and Delhomme, 2005; Parker et al., 1992; Stasson and Fishbein, 1990; Wallén Warner and Åberg, 2008).

Much of the literature in which the TPB is applied does not follow the formal structure of the theory, but rather 'extends' the theory by adding additional explanatory factors, often to better deal with the issue of emotional content that practitioners believe is not adequately incorporated in the original model (Parker et al., 1995; Stradling and Parker, 1997). Additional modelling approaches (Assailly, 2011; and Gerrard et al., 2008) have been proposed to better incorporate non-voluntary and non-planned behaviour. The model reported in this paper explicitly includes a factor to represent emotional content and a factor which reflects the possibly non-rational assessment of risk.

Over the past two decades the concept that the personality trait of *sensation seeking* is associated with risky driving behaviour has been well established (Brown and Cotton, 2003; Horvath and Zuckerman, 1992; Zuckerman, 1991, 1994, 2006) and may explain 10–15 per cent of risky driving (Jonah, 1997). People with this trait consider risk taking as an end in itself. For this group of drivers the motive behind risky driving is the thrill associated with risk taking. Zuckerman (1994) defined sensation seeking as follows: 'Sensation seeking is a *trait* defined by the seeking of varied, novel, complex, and intense sensations and experiences, and the willingness to take physical, social, legal, and financial risks for the sake of such experience.' Thrill-seekers rated speeding as more exciting than those who scored low on the same thrill-seeking scale (Jonah, 2001). Thrill seekers have been linked to risky driving habits and high involvement in road crashes (Farley, 1991; Singh, 1992; Thorson and Powell, 1987). Stephenson et al. (2003) examined 'brief measures' for screening, which influenced our decision to reflect a combination of intensity and novelty only (see also Arnett, 1994).

A second major theme in the literature of the social psychology of driving behaviour is the concept of failure of *perception of risk* in driving behaviour (Brown and Cotton, 2003; Iversen, 2004; Ulleberg and Rundmo, 2003). Machin and Sankey (2008) hypothesized that 'a higher level of perceived risk for a particular behaviour is associated with a lower chance that an individual would take part in that behaviour'. Other studies have focused more on drivers' perception of the consequences of their behaviour and argued that drivers do not actively worry about the disastrous outcomes of risky driving behaviour, but rather tend to focus on the positive aspects of speeding (Horswill and McKenna, 1999; Stradling et al., 1997). Although risky drivers are aware of the risks involved in violating the speed limit, they do not believe that they themselves are at risk (Forward, 2006). It would then appear that risky drivers hold beliefs that minimize the perception of personal risk. This was also confirmed by Ben-Zur and Reshef-Kfir (2003) who in addition to perception of benefits found that the use of avoidance-coping strategies, including denial, determined the decision to take risks (see also Rosenbloom, 2003). In a similar manner, one can deny (or downplay) the idea that there will be later experiences of regret; a study by Parker et al. (1995) included the concept of 'anticipated regret' which can be described as the 'value of consequences' (see Triandis, 1980).

The systematic comparison of urban and rural driving and crash patterns has received relatively little attention until recently. Recent contributions have been made by Blatt and Furman (1998), Rakauskas et al. (2009), and Ward (2007) in the United States, and several contributions come from the DRIVE Study in Australia, including Chen et al. (2009, 2010). Rakauskas et al. reported that, 'rural respondents had significantly lower sensation seeking tendencies and had a lower perceived utility of enforcement and engineering' interventions; their rural group had a lower perception of the 'dangerousness' from not using seat belts. In a rural study in Australia, Sticher (2005) found that 'rural road users who inaccurately appraised the risk factors associated with rural crashes gave very positive appraisals of their own driving ability and incorrectly attributed risk to external factors'.

Method

We undertook a survey of over 1,000 rural and non-rural residents in three US states. The sampling procedure was undertaken with the goal of getting a stable population for the final examination of the rural versus non-rural differences for males under 30 years old. There were 998 valid participants, of which 468 were rural, and 530 were non-rural; 605 were male and 393 were female; 423 were under 30, and 575 were aged 30 years and older.

Our survey instrument was designed to test a wide variety of theories concerning the relationship of attitude to risky driving behaviour. The questions were designed to elicit reactions concerning objective norm, injunctive norm, affective attitude, instrumental attitude, denial of danger, optimism bias, anticipated regret, overall sensation seeking, intensity of sensation seeking, novelty in sensation, social capital, mildly deviant anti-social behaviour, trust/altruism, beliefs about alcohol, annoyance/impatience, self-efficacy, and dissatisfaction with personal conditions. The survey also collected data on outcome behaviours including speed oriented activities, seatbelt use, alcohol driving, and other errant driving behaviours. Factors relevant to risky driving behaviour were examined through several methods, including exploratory factor analysis and hierarchical multiple regression applied to the results of that process.

Step One: Creation of the Measurement Model

Consistent with a 'two-step' process advocated in the SEM literature (Kline, 2005), a measurement model was examined separately from the structural regression model. The measurement model (Figure 2.1) created using the AMOS 16 software program (Arbuckle, 2007), has five latent factors, each of which is based on two indicators representing a multi-item summed scale. During the process of confirmatory factor analysis, each observed indicator was found to load onto one, and only one, latent variable. To confirm this, each observed indicator was tested to see if it would better load on another latent factor, or, for any strong loading factors, on second or third latent variables. In no case did this happen, and the discriminant nature of the separate factors was confirmed (Brown, 2006). Correlations between all pairs of latent factors (expressed as two directional arrows in SEM graphics) were examined, with all correlations well under .90, which again confirms that each latent factor is independent of others. At the same time, all covariances among latent factors are statistically significant at $p \leq .001$. The loadings (expressed as one directional arrows in SEM graphics) between the latent factors and the observed indicator are satisfactory, with only one loading below .6.

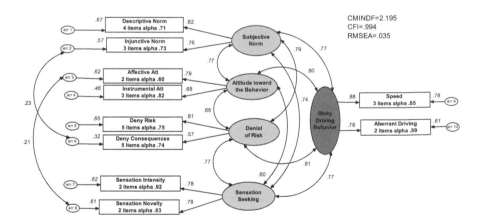

Figure 2.1   The Measurement Model. Definitions: CMINDF = chi square divided by degrees of freedom; CFI= comparative fit index; RMSEA= root mean square error of approximation. All coefficients and correlations significant at $p \leq .00.1$

The measurement model resulting from the Confirmatory Factor Analysis includes four unobserved latent factors to aid in the explanation of risky driving behaviour. The four latent factors are:

1. '*Subjective Norm*' a latent factor which is associated with two manifest observations. The indicator 'Descriptive Norm' is a summed four-item scale from the inclusion of: 'People who are important to me will drive over 45 mph through such towns in the next two months', 'The people I like to be around would never obey these town speed limits'; 'The people I like to be around will tailgate the slow car in the next two months', 'The friends I hang out with think it is fun to see how fast you can go'. Together these four items have a Cronbach's *alpha* measure of internal consistency of .71. The observed indicator 'Injunctive Norm' is a three-item summed scale from inclusion of: 'If I told my friends at dinner that I broke the speed limit to get home, they would ... approve'; 'My parents would find the idea of me speeding ... acceptable'; 'If I told most of the people whose opinion I care about that I tailgated the car, they would ... approve'. Together these three items have a Cronbach's alpha of .74.

2. '*Attitude toward the Behaviour*' a latent factor associated with two manifest observations. The indicator Instrumental Attitude is a summed three-item scale from the inclusion of: 'Driving over 45 mph through the towns would help to get me home in time for dinner'; 'Speeding through the towns would allow me to arrive home much sooner'; and 'Driving at the higher speed in the passing lane would get me home a lot faster'. Together these three items have a Cronbach's alpha measure of internal consistency of .82. The indicator Affective Attitude is a summed two-item scale from the inclusion of: 'I would feel really annoyed if I had to drive behind such a slow-moving vehicle'; and 'It would be easier for me to follow the speed laws if I wasn't so impatient'. Together these three items have a Cronbach's alpha of .60.

3. '*Denial of Risk*' a latent factor associated with two manifest observations. The indicator Deny Risk is a summed five-item scale from the inclusion of: 'Speed limits do not save lives'; 'The risk of dying in a traffic crash is so low that you can ignore it'; 'When I speed, I'm only putting myself in danger, not others'; 'I have a very good car that is safe to drive considerably above the speed limit'; and 'Driving after having a few drinks is acceptable'. Together these five items have a Cronbach's alpha measure of internal consistency of .75. The indicator 'Deny Consequences' is a summed five-item scale from the inclusion of reversed items, resulting in variables which report: 'I would (not) feel ashamed to be pulled over by the police for speeding through these towns'; ' I would (not) feel ashamed to be pulled over by the police for tailgating the slow car'; 'Going faster

through the towns (would not) make me feel nervous'; 'Driving close to the car in front of me would (not) make me nervous'; 'Hurting someone else with my car would (not) scar me for life'. Together these five items have a Cronbach's alpha of .74.

4. 'Sensation Seeking' a latent factor associated with two manifest observations. The indicator Sensation Seeking: Intensity is a summed two-item scale from the inclusion of 'How often do you do exciting things, even if they are dangerous'; and 'How often do you do dangerous things for fun?'. Together these two items have a Cronbach's alpha measure of internal consistency of. 92. The indicator Sensation Seeking: Novelty is a summed two-item scale: 'I prefer friends who are exciting and unpredictable' and 'I like new and exciting experiences, even if I have to break the rules'. Together two items have a Cronbach's *alpha* of .63.

The outcome latent factor 'Risky Driving Behaviour' is associated with two manifest observations, with all questions drawn from the Driver Behaviour Questionnaire (Parker et al., 1995). The 'Indicator Speed' is a summed three-item scale from the inclusion of 'Go more than 80 mph on a rural interstate', 'Disregard the speed on speed limit on a two-lane highway' and 'Disregard the speed limit on a residential road'. The indicator 'Aberrant Driving' is a summed two-item scale from inclusion of 'Race away from the traffic lights with the intention of beating the driver next to you', and 'Pass a slow driver on the right'. Together these two items have a Cronbach's alpha measure of .59.

The measurement model for the project is well fitting. Root mean square error of approximation (RMSEA) is .035, below the desired maximum value of .05, with a 90 per cent confidence level that the high range of the value would be under .048. The comparative fit index (CFI) is .99, well above the desired minimum value of .95 (Schumacker and Lomax, 2004). The widely used index of chi-square divided by the degrees of freedom is 2.2, well below the desired upper range of 5.0; the use of this index has been widely challenged in the literature but is reported here for consistency with the traditional use of chi-square reporting (Kline, 2005).

The measurement model includes two correlations of error measurements, both of which were suggested in the model modification process of AMOS. The two correlations were found to be consistent with theory, in which similarities between the content of variables (survey questions) could explain the logic of allowing the correlation of error terms. Both the correlations accepted in the model reflected covariances which were statistically significant at $p \leq .001$.

Step Two: Creation of the Structural Regression Model

Having established that the latent factors do in fact measure what we intend them to measure, and that the measurement model has an appropriate goodness of fit, and is consistent with theory, the next step focused on the manner in which the latent factors interact with each other in the context of simultaneous regression analyses for multiple samples.

The structural component of the full model was created by a process of model refinement. The initial model hypothesized Attitude toward the Behaviour, Subjective Norm and Denial of Risk (used in place of Perceived Behavioural Control) would directly explain the outcome latent factor, with any additional factors such as personality traits influencing behaviour only through one or more mediating factors, generally following the underlying theory.

In assessing how the factors from the Theory of Planned Behaviour interrelate with the additional factors, several alternative structural models were analysed (see also Chan et al., 2010). It was determined that the coefficient from Subjective Norm and Risky Driving Behaviour was statistically insignificant, and that parameter was dropped from the model. At the same time, it was determined that the coefficient from Sensation Seeking to Risky Driving Behaviour was significant, and its inclusion improved overall model fit (by all measures.) Thus, in the model (Figure 2.2 and Table 2.1), Risky Driving Behaviour is directly explained by three latent factors, Attitude toward the Behaviour, Denial of Risk, and Sensation Seeking; it is also indirectly explained by Subjective Norm, Sensation Seeking, and Denial of Risk. Additionally, a new parameter from Denial of Risk to Attitude

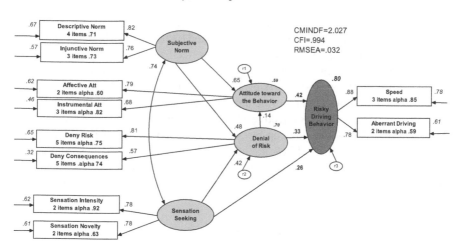

Figure 2.2   **The Full SEM Model, including measurement and structural regression components, shown with standardized coefficients. All coefficients significant at $p \leq .001$, except denial to attitude, at $p \leq .08$**

Table 2.1   Coefficients for the Structural Regression Model (Full Sample, n= 998)

			Unstandardized			Standardized
			Estimate	S.E.	P	Estimate
Denial of Risk	<---	Subjective Norm	0.437	0.059	***	0.478
Denial of Risk	<---	Sensation Seeking	0.374	0.057	***	0.42
Attitude toward the Behaviour	<---	Subjective Norm	0.664	0.083	***	0.653
Attitude toward the Behaviour	<---	Denial of Risk	0.16	0.09	0.076	0.143
Risky Driving Behaviour	<---	Sensation Seeking	0.292	0.067	***	0.265
Risky Driving Behaviour	<---	Denial of Risk	0.408	0.093	***	0.329
Risky Driving Behaviour	<---	Attitude toward the Behaviour	0.469	0.055	***	0.421

toward the Behaviour was also found to improve overall model fit. Its inclusion in the SEM model for the full sample is optional, as it was significant only at $p \leq .08$; however, the parameter proves to be significant for the rural population (discussed below) and thus has not been deleted from the overall model applied to the full sample.

The relative strength of the total influence of each of the explanatory factors is presented in the SEM calculation of 'standardized total effect', which is calculated by adding the direct effect with the indirect effect. As described in the output of the AMOS programmme, the standardized total (direct and indirect) effect of Subjective Norm on Risky Driving Behaviour is .46. That is, due to both the direct (unmediated) and indirect (mediated) effects of Subjective Norm on Risky Driving Behaviour, when Subjective Norm goes up by one standard deviation, Risky Driving Behaviour goes up by .46 standard deviations (Arbuckle, 2007). In spite of the fact that its effects are all mediated by other variables, Subjective Norm emerges as the largest generator of total effect, at .46, compared with .43 for Sensation Seeking, .42 for Attitude toward the Behaviour, and .39 for Denial of Risk.

Undertaking the Urban versus Rural Analysis

The full sample was then divided into a rural group, and a non-rural group, including those who describe themselves as urban and suburban. For purposes of brevity, we have labelled the second group as 'urban' even though it is most accurately referred to as 'other than rural'. We used the technique of multi-sample modelling, in which the structural equation modelling software analyses two samples simultaneously. From this point on, the paper examines the *difference* between the patterns of the two population groups in terms of the factors which influence their propensity to undertake risky driving behaviour.

The model was deemed an adequate medium through which to observe rural versus non-rural differences. When the model is run with both samples simultaneously, a high level of goodness of fit results; with an overall CFI of .99 and a RMSEA of .02. The measure of chi-square divided by the degrees of freedom, at 1.7 (less than the upper limit of 5.0 desired) is another indication that the model fits the data extremely well.

Results

Figure 2.3 shows the path diagrams for the two structural regression models, representing the rural and urban groups. Looking at the direct parameters (expressed as unstandardized coefficients for the comparison of groups), the application of the model to the full population included in our survey shows a marked difference between the roles of the latent factors in the explanation of risky driving behaviour. Specifically, the unstandardized coefficient for the parameter between Sensation Seeking and Risky Driving Behaviour is .50 for the urban group, while it is only .08 (which is statistically insignificant at $p \leq .05$) for the rural group. In a parallel vein, the coefficient from Denial of Risk to Risky Driving Behaviour is .56 for the rural group, while only .26 for the urban group.

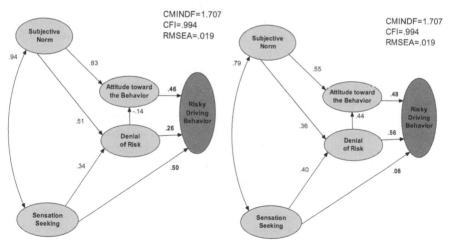

Figure 2.3   **Path Diagrams for Urban (left) and Rural Models (right), with unstandardized coefficients. All significant at $p \leq .001$, except denial to attitude in the urban model, and sensation to risky driving in the rural model**

A more inclusive description of the roles of the factors in the influence of risky driving behaviour is provided from standardized total effects: in the urban group, the standardized total effect for Sensation Seeking is significantly higher than that for Denial of Risk (at .57 versus .20), with the opposite pattern for the rural group: in the rural group, the total effect for Denial of Risk is higher than that for Sensation Seeking (at .77 versus .39).

We undertook the rural versus urban comparison of the roles of the four latent variables, with particular emphasis on the interplay between Sensation Seeking and Denial of Risk, for six populations: (1) the total sample, (2) all the males, (3) all the females, (4) all those under 30, (5) all those over 30, and (6) all males under 30. For all six of these populations, the standardized total effect for Denial of Risk was more important than that for Sensation Seeking in the rural segment and *less* important in the urban segment.

Denial of Risk and Sensation Seeking, by Geographic Group

Table 2.2 presents the model output for the total standardized effect of each of the four latent explanatory factors on Risky Driving Behaviour. For emphasis, we have shaded each of the values for Sensation Seeking for the urban group and Denial of Risk for the rural group. For each of the six populations tested Table 2.2 shows Sensation Seeking to be a stronger factor than Denial of Risk for the urban group, with Denial of Risk the stronger factor for the rural group. Importantly, this pattern, established in the full sample, is confirmed in the highest risk population group, young males.

For the sample groups containing males and younger participants, Sensation Seeking is the strongest of the four factors in the urban context, with Denial of Risk the strongest of the four factors in the rural context. Table 2.2 shows that for the female sample and for the older sample, Sensation Seeking is less important than for the males and younger survey participants for both geographic groups. While the rural group in the female and older samples does continue to show a major role for Denial of Risk, this role is shared with significant roles from the other latent factors, including Subjective Norm and Attitude toward the Behaviour. In general, the results shown in Table 2.2 are consistent with established literature that Sensation Seeking is most relevant to males and younger persons.

Figure 2.3 shows that the directional arrow running from Denial of Risk to Attitude toward the Behaviour is statistically significant for the rural group, but is not for the urban group. It seems that the power of the Denial of Risk factor in the rural groups might well influence the formation of attitude towards the basic desirability of speeding. While causality is not asserted, holding a view that speeding has little risk or downside consequences might influence one's basic evaluation of the goodness of that speeding. On the other hand, this pattern may not apply for the urban population.

Table 2.2   Standardized total effect on risky driving from the four factors

	Social Norm	Attitude toward the behaviour	Denial of risk	Sensation seeking
Full sample				
Urban	0.42	0.38	0.16	0.50
Rural	0.49	0.48	0.62	0.36
Males only				
Urban	0.36	0.27	0.12	0.54
Rural	0.55	0.45	0.65	0.28
Young only				
Urban	0.37	0.33	0.10	0.59
Rural	0.41	0.41	0.78	0.40
Young males only				
Urban	0.35	0.33	-0.02	0.60
Rural	0.50	0.49	0.78	0.29
Female only				
Urban	0.50	0.63	0.18	0.33
Rural	0.43	0.53	0.55	0.44
Older only				
Urban	0.48	0.54	0.15	0.33
Rural	0.54	0.61	0.54	0.27

Evidently, for the urban populations surveyed in the project, the need for sensational experiences may be so strong that the question of evaluating the downside risk associated with those experiences becomes relatively less important (see also Rosenbloom, 2003). The rural population seems to be less directly affected by the sensation seeking personality trait, but seems to deny the seriousness of the downside risk of their risky driving behaviour when undertaking it. In the explanation of risky driving behaviour, a clear pattern has emerged in which bad judgements about the implications of speed and speed related behaviours are associated with the denial and downplaying of risk. Table 2.2 shows that, for the rural population, this denial pattern better explains the behaviour than does the direct impact of the Sensation Seeking personality trait – a pattern true for all six rural groups examined.

Mean Differences between Urban and Rural Groups

In addition to studying the patterns of variance through the application of regression analysis within SEM, we examined the difference in mean values for the 10 indicator variables. Two of those are loaded on the latent factor Risky Driving Behaviour and eight are loaded on the four explanatory latent factors, see Figure 2.1. The mean values help establish some of the most basic observations about differences in driving behaviour between the two groups in our sample.

First, generalizations can be made about the urban versus rural differences in self-reported driving behaviour. Looking at Table 2.3, the urban group has worse self-reported speed and aberrant behaviour than the rural group, with both differences being significant. The pattern of the urban group having higher levels of risky driving patterns is consistent with previous work with this data (Coogan, 2009; Coogan et al., 2011) and with other data (Rakauskas et al., 2009) and is generally consistent with the fact that urban drivers have a higher number of crashes than do rural drivers.

Second, for each of the eight indicator measures created in the confirmatory factor analysis process the urban group holds a set of attitudes, beliefs, and personality trait characteristics which are more problematic than those of the rural group. Table 2.4 shows that it is clearly inaccurate to look at the rural drivers as more problematic than the urban drivers, and that the differences between the two groups require further analysis. In terms of the mean values of the two geographic groups, the urban group scores higher than the rural group on both indicators concerning Sensation Seeking, which is consistent with the literature. Less clear is the fact that the urban group also scores higher on both Deny Risk and Deny Consequences, which seems to be inconsistent with the direction established in the analysis of variance through the regressions included in the SEM process.

The data suggest that the while both groups have members who deny both the fact of risk and the downside consequences of that risk, this pattern is far stronger in explaining risky driving behaviour in the rural context than in the urban context. Similarly, while both geographic groups have members who display the patterns of the sensation seeking personality, the presence of such a personality trait is far stronger in explaining risky driving behaviour in the urban context than in the rural context. Both of these relationships should be examined in further research.

Limitations and Further Research

The survey sample was created as part of broader study of rural issues, and covers three states in the United States which are each predominantly rural in nature. Thus, the non-urban sample used in this study is based on the smaller urban areas

Table 2.3   Rural/urban differences in mean values for self-reported speed-related driving behaviour

Group	Speed 3 items *alpha* .85	Aberrant Driving 2 items *alpha* .59
Urban sample, n = 530	2.92	2.72
Rural sample, n = 468	2.63	2.31
Total, n = 998	2.78	2.52

Differences significant at $p \leq .01$

Table 2.4   Rural/urban differences in mean values for the eight explanatory indicators

Group	Descriptive Norm 4 items alpha .71	Injunctive Norm 3 items alpha .73	Affective Attitude 2 items alpha .60	Instrumental Attitude 3 items alpha .82	Deny Risk 5 items alpha .75	Deny Consequence 5 items alpha .74	Sensation Intensity 2 items alpha .92	Sensation Novelty 2 items alpha .63
Urban sample n = 530	3.23	3.12	3.94	4.31	2.62	2.54	2.73	3.59
Rural sample = 468	2.97	2.90	3.57	3.87	2.33	2.38	2.43	3.22
Total	3.11	3.02	3.77	4.10	2.49	2.47	2.59	3.42

All of the urban/rural differences in mean values are statistically significant at $p \leq .001$, except Deny Consequences at $p \leq .05$

located in the three states, and larger suburban population. The next phases of this study will include more urban participants from states not generally characterized as rural. It should also be noted that the sample was structured to support analysis of the most problematic drivers (meaning males and younger persons); for each of the six populations analysed the sample supports an analysis of urban versus rural participants, which is the central purpose of the research. The sample was not meant to represent the driving population as a whole, but rather the groups which are the central focus of the research. Finally, the data are cross sectional in nature, limiting the interpretation of causality.

This paper has focused on the differences between urban and rural populations in the manner in which they undertake risky driving behaviour. The research implies that the rural group, and specifically the younger male members of that group, may have developed a pattern of underestimation of the downside implications of risky driving behaviour. Further research will be needed to determine *why* these differences exist, and how to deal with them in the design of social interventions and other strategies, which were beyond the scope of this exercise.

Conclusions

These results should be retested in a larger sample, and one which includes more participation from both the suburban and urban segments, which should be considered separately. If such a larger sample confirms our observations, then strategies to deal with the rural segment of the population should focus on the evident denial of downside risk. For urban segments policy should deal with the significant problem of a personality trait condition influencing the undertaking of risky driving behaviour. An improved understanding of the phenomenon of denial of risk noted for the rural populations in this study could have implications for other domains of public health research.

References

Ajzen, I. (1991). The theory of planned behavior. *Organizational Behavior and Human Decision Processes, 50*, 170–211.

Arbuckle, J. (2007). *Amos 16.0 Users Guide*, Amos Development Corporation, distributed by SPSS, Inc.

Arnett, J.J. (1994). Sensation seeking: A new conceptualization and a new scale. *Personality and Individual Differences, 16*(2), 289–96.

Assailly, J.P. (2011). *The Psychology of Risk*. New York: Nova Science (forthcoming).

Bamberg, S., Hunecke, M., and Blobaum, A. (2007). Social context, personal norms and the use of public transportation: Two field studies. *Journal of Environmental Psychology, 27*, 190–203.

Ben-Zur, H., and Reshef-Kfir, Y. (2003). Risk taking and coping strategies among Israeli adolescents. *Journal of Adolescence, 26*(3), 255–65.

Blatt, J., and Furman, S.M. (1998). Residence location of drivers involved in fatal crashes. *Accident Analysis and Prevention, 30*(6), 705–11.

Brown, S.L., and Cotton, A. (2003). Risk-mitigating beliefs, risk estimates and self-reported speeding in a sample of Australian drivers. *Journal of Safety Research, 34*, 183–8.

Brown, T.A. (2006). *Confirmatory Factor Analysis for Applied Research*. New York: The Guilford Press.

Budd, R.J., North, D., and Spencer, C. (1984). Understanding seat-belt use: A test of Bentler and Speckart's extension of the 'theory of reasoned action'. *European Journal of Social Psychology, 14*, 69–78.

Chan, D.C.N., Wu, A.M.S., and Hung, E.P.W. (2010). Invulnerability and the intention to drink and drive: An application of the theory of planned behavior. *Accident Analysis and Prevention, 42*, 1549–55.

Chen, H.Y., Ivers, R.Q., Martiniuk, A.L.C., Boufous, S., Senserrick, T., Woodward, M., Stevenson, M., Williamson, A., and Norton, R. (2009). Risk and type of crash among young drivers by rurality of residence: Findings from the DRIVE study. *Accident Analysis and Prevention, 41*, 676–82.

Chen, H.Y., Ivers, R.Q., Martiniuk, A.L.C., Boufous, S., Senserrick, T., Woodward, M., Stevenson, M., and Norton, R. (2010). Socioeconomic status and risk of car crash injury, independent of place of residence and driving exposure: Results from the DRIVE study. *Journal of Epidemiological Community Health, 64*, 998–1000.

Coogan, M. (2009). *Looking for a Rural Culture of Driving*. Presentation to the Center for Excellence in Rural Transportation Safety, University of Minnesota, Williamsburg, VA.

Coogan, M., Campbell, M., Adler, T., Forward S., and Assailly, J.P. (2011). Latent Class Cluster Analysis of Driver Attitudes Towards Risky Driving in Northern New England: Is There a Rural Culture of Unsafe Driving Attitudes and Behavior? Presentation to the *90th Annual Meeting of the Transportation Research Board*, Washington DC. Accessible at TRB Website.

Elliott, M.A., Armitage, C.J., and Baughan, C.J. (2003). Drivers' compliance with speed limits: an application of the theory of planned behavior. *Journal of Applied Psychology, 88*, 964–72.

Elliott, M.A., Armitage, C.J., and Baughan, C.J. (2005). Exploring the beliefs underpinning drivers' intentions to comply with speed limits. *Transportation Research Part F: Traffic Psychology and Behaviour, 8*, 459–79.

Farley, F. (1991). The type-t personality. In L.P. Lipsitt and L.L. Mitnick. (Eds.). *Self-regulatory Behavior and Risk Taking: Causes and Consequences*. Norwood, NJ: Ablex, pp. 371–82.

Fishbein, M., and Ajzen, I. (2010). *Predicting and Changing Behavior: The Reasoned Action Approach*. New York: Psychology Press.

Forward, S. (2006). The intention to commit driving violations: A qualitative study. *Transportation Research Part F: Traffic Psychology and Behaviour*, *9*, 412–26.

Forward, S. (2009). The theory of planned behaviour: The role of descriptive norms and past behaviour in the prediction of drivers' intention to violate. *Transportation Research Part F: Traffic Psychology and Behaviour, 12*, 198–207.

Forward, S. (2010). Intention to speed in a rural area: Reasoned but not reasonable. *Transportation Research Part F: Traffic Psychology and Behaviour, 13*, 223–32.

Gerrard, M., Gibbons, F., Houlihana, A., Stocka, M., and Pomerya, E. (2008). A dual-process approach to health risk decision-making: The prototype willingness model, *Developmental Review, 28*, 29–61.

Horswill, M.S., and McKenna F.P. (1999). The effect of perceived control on risk taking. *Journal of Applied Social Psychology, 29*, 377–91.

Horvath, P., and Zuckerman, M. (1992). Sensation seeking, risk appraisal and risky behaviour. *Personality and Individual Differences, 14*(1), 41–52.

Iversen, H. (2004). Risk-taking attitudes and risky driving behaviour. *Transportation Research Part F: Traffic Psychology and Behaviour, 7*, 135–50.

Jonah, B.A. (1997). Sensation seeking and risky driving: A review and synthesis of the literature. *Accident Analysis and Prevention, 29*(5), 651–65.

Jonah, B.A., Thiessen, R., and Au-Yeung, E. (2001). Sensation seeking, risky driving and behavioural adaptation. *Accident Analysis and Prevention, 33*, 679–84.

Kline, R.B. (2005). *Principles and Practice of Structural Equation Modeling*. New York: The Guilford Press.

Letirand, F., and Delhomme, P. (2005). Speed behaviour as a choice between observing and exceeding the speed limit. *Transportation Research Part F: Traffic Psychology and Behaviour, 8*, 481–92.

Machin, M.A., and Sankey, K.S. (2008). Relationships between young drivers' personality characteristics, risk perceptions and driving behavior. *Accident Analysis and Prevention, 40*(2), 541–7.

New England Transportation Institute (NETI) (2010). *Final Report to the Federal Highway Administration describing grant activities conducted under contract DTF61-06-00008, Volume Two: Safety and Information*, Chapter One, p. 2. Based on research undertaken by SmartMobility, Inc.

NHTSA. *FARS Data Tables*. http://www-fars.nhtsa.dot.gov/Main/index.aspx. Accessed December 2010.

National Center for Health Statistics, National Vital Statistics System, Office of Statistics and Programming. *10 leading causes of death*. Available from http://webappa.cdc.gov/cgi. Accessed Oct 30, 2010.

Parker, D., Manstead, A.S.R., Stradling, S.G., and Reason, J.T. (1992). Determinants of intention to commit driving violations. *Accident Analysis and Prevention, 24*, 117–31.

Parker, D., Reason, J.T., Manstead, A.S.R., and Stradling, S.G (1995). Driving errors, driving violations and accident involvement. *Ergonomics, 38*, 1036–48.

Parker, D., Manstead, A.S.R., and Stradling S.G. (1995). Extending the theory of planned behaviour: the role of personal norm. *British Journal of Social Psychology, 34*, 127–37.

Rakauskas, M.E., Ward N.J., and Gerberich, S.G. (2009). Identification of differences between rural and urban safety cultures. *Accident Analysis and Prevention, 41*, 931–7.

Rosenbloom, T. (2003). Risk evaluation and risky behavior of high and low sensation seekers. *Social Behavior and Personality, 31*(4), 375–86.

Schumacker, R.E., and Lomax, R.G. (2004). *A Beginner's Guide to Structural Equation Modeling.* Mahwah, NJ: Lawrence Erlbaum Associates, Inc. Publishers.

Singh, A. (1992). Alcohol, driving and young people. Paper presented at *Eurosafe 1992 – Safer Driving in Europe.* London, UK: Association of London Borough Road Safety Officers.

Stasson, M., and Fishbein M. (1990). The relation between perceived risk and preventive action: A within-subject analysis of perceived driving risk and intentions to wear seat-belts. *Journal of Applied Social Psychology, 20*, 1541–57.

Stephenson, M.T., Hoyle, R.H., Palmgreen, P., and Slater, M.D. (2003). Brief measures of sensation seeking for screening and large-scale surveys. *Drug and Alcohol Dependence, 72,* 279–86.

Sticher, G. (2005). An investigation of attitudes towards risk factors, personal driving ability and road safety information among rural and remote users. Proceedings *Conference of Australian Institutes of Transport Research,* Brisbane, Australia.

Stradling, S.G., and Parker, D. (1997). Extending the theory of planned behaviour: The role of personal norm, instrumental beliefs and affective beliefs in predicting driving violations. In T. Rothengatter and V. Carbonell (Eds.), *Traffic and Transport Psychology: Theory and Application.* New York: Pergamon Press.

Thorson, J.A., and Powell, F.C. (1987). Factor structure of a lethal behaviours scale. *Psychological Reports, 61*, 807–10.

Triandis, H.C. (1980). Values, attitudes and interpersonal behaviour. In H. Howe and M. Page (Eds.), *Nebraska Symposium on Motivation.* Lincoln, NE: University of Nebraska Press, pp. 195–259.

Ulleberg, P., and Rundmo, T. (2003). Personality, attitudes and risk perception as predictors of risky driving behaviour among young drivers. *Safety Science, 41*, 427–43.

Wallén Warner, H., and Åberg, L. (2008). Drivers' beliefs about exceeding the speed limits. *Transportation Research Part F: Traffic Psychology and Behaviour, 11*, 376–89.

Ward, N. (2007). The culture of traffic safety in rural America. In *Improving Traffic Safety Culture in the United States The Journey Forward*. Washington, DC: AAA Foundation for Traffic Safety, pp. 149–6.

Zuckerman, M. (1991). Sensation seeking: the balance between risk and reward. In L.P. Lipsitt and L.L. Mitnick (Eds.), *Self-regulatory Behaviour and Risk Taking: Causes and Consequences*. Norwood, NJ: Ablex, pp. 143–52.

Zuckerman, M. (1994). *Behavioral Expressions and Biosocial Bases of Sensation Seeking*. New York, USA: Cambridge University Press.

Zuckerman, M. (2006). *Sensation Seeking and Risky Behavior*. Washington DC: American Psychological Association.

Chapter 3

Executive Function Development and Stress Effects on Driving Performance: Preliminary Findings from a Young Adult Sample

Melanie J. White, Ross McD. Young and Andry Rakotonirainy
Centre for Accident Research and Road Safety, Institute of Health and Biomedical Innovation, Queensland University of Technology, Australia

Introduction

This chapter presents a pilot study examining the interactive contributions of executive function development/impairment and psychosocial stress to young adults' (17–30 years old) driving behaviour in a simulator city scenario.

Learning to drive is an exciting but challenging experience for young people. This period is an important focus for road safety policy and research. However, it is also a time at which young adults' brains, particularly in frontal areas important for higher order cognitive abilities, are still developing (Blakemore and Choudhury, 2006). Driving is a complex task requiring the integration of several of these higher-order cognitive functions, such as working memory, sustained attention, attention shifting, interference control, risk perception and response inhibition (both risk taking and the ability to suppress a habitual response in a novel situation) (Barkley et al., 2002; Wickens et al., 2008). Collectively, these cognitive abilities are referred to as executive function.

Transport crashes are the leading cause of death among young people aged 15–24 years in Australia (32 per cent of all deaths of young people in 2004) (Australian Institute of Health and Welfare, 2007). Young novice drivers are over-represented by two to four times that of other drivers in transport accidents in Australia and internationally (Smart et al., 2005). Developmental impairments in executive function associated with adolescence could potentially contribute to this. A recent study reported negative associations between self-reported executive function and self-reported driving behaviours including driving violations, mistakes and lapses (Morris and Dawson, 2008). This study also found some evidence of younger drivers with poorer executive function being more likely to demonstrate these negative driving behaviours (using the total scaled score) than their peers without executive dysfunction. Findings from a recent qualitative study similarly suggest that driving-related skill deficits and risky driving behaviours (and their motivations) in young learner drivers with Attention Deficit Hyperactivity Disorder

(ADHD) are related to the core executive function impairments characterized by the disorder (White, 2009). However, further research is needed using more objective measurement methods to confirm the relationships between executive function and driving behaviours in young people and to examine the mechanisms underlying the association.

Executive function primarily involves dopaminergic pathways in the brain (Goldman-Rakic et al., 2000; Mitchell and Phillips, 2007). Acute psychosocial stress has been found to increase dopamine neurotransmission in the ventral striatum of humans (Pruessner, 2004). Previous research has showed that genes which are associated with fewer dopamine receptors in this part of the brain (DRD2 C957T, ANKK1 *Taq*1A) are associated with differential aspects of executive function and impulsivity. Further, some of these relationships are moderated by acute and chronic forms of psychosocial stress exposure (possibly due to stress-induced dopamine release mechanisms) (White et al., 2008; 2009). Individual physiological responsiveness to stress (and therefore greater vulnerability to its effects) may in part be due to these genetic risk factors and lead to worsening executive function.

The current research was designed as a pilot study to test the feasibility of the methodology for a larger study. It examined the interactive contributions of psychosocial stress and executive function development or impairment to young adults' driving behaviour in a driving simulator.

Method

Participants

A sample of 40 (9 male) right-handed young adults aged 17–30 years (M = 21.56 years, SD = 2.99), with at least six months licensed driving experience and good/ corrected vision were recruited from the Queensland University of Technology and local community via advertising. Additional exclusion criteria, elicited via self-report, included health problems or medications affecting the endocrine system, current psychiatric illness, moderate to heavy caffeine intake (more than 300 mg/ day) or alcohol intake (more than two standard drinks/day), and any current illicit drug use, primarily to avoid confounds with the cortisol analyses. Participants were offered either $30 or course credit (in the case of psychology students) as a token of appreciation and in recognition of their time.

Procedure and Measures

The full experimental protocol is provided here, however only those selected measures for which preliminary data are presented in this chapter are described in detail. Participants were randomly allocated to either an acute stress or relaxation induction condition and were tested individually in a laboratory on the university campus, in two hour sessions held between 12 and 6 p.m. (to control for variations of cortisol levels over the circadian rhythm; Fan et al., 2009). In accordance with the Salivette® saliva cortisol testing protocol, participants were asked not to eat or drink 30 minutes prior to their arrival. After providing informed consent, participants were fitted with EEG and ECG electrodes to their scalp and chest respectively, which remained attached for the duration of the experiment. Participants provided a saliva sample for DNA extraction and completed a pre-induction battery of questionnaires, which included questions about demographics, personality, chronic stress, general health and driving behaviour. Participants completed four performance-based measures of executive function. These included, in order, the paper-and-pencil timed Trail Making Test (TMT) Parts A and B, with Part B assessing set-shifting ability (Spreen and Strauss, 1998), and three computerized tasks assessing stop inhibition, working memory and emotional processing: the GoStop stop signal task (Dougherty et al., 2005; see also White et al., 2008), a digit 'identity match' *n*-back paradigm at n = 1, 2, and 3 levels (Knops et al., 2006) and an emotional faces go/no-go task (Sackler Institute for Developmental Psychobiology, 2007), respectively.

   Following a five-minute stress or rest induction procedure (see White et al., 2008 for details), participants then completed two desktop driving simulator sessions of approximately 7–10 minutes each (Brisbane Central Business District city scenario, followed by a Cairns Captain Cook Highway scenario which is not reported on here). Scenarios were developed using the SCANeR software (OKTAL, France). Pre-induction and post-induction samples of saliva were taken via Salivette® (SARSTEDT, Germany) for cortisol analysis. Self-report visual analogue scale ratings of subjective feelings of stress were also collected at each time period as a test of the validity of the stress manipulation.

CBD Driving Simulator Task

Participants were seated in front of a steering wheel with accelerator and brake pedals at their feet. The driving scene was projected onto a screen (128 cm × 145 cm) in front of the participant with the speedometer projected at the bottom centre of the screen and the rear-view mirror at the top centre. Participants drove two laps around a computer modelled scenario of the Brisbane Central Business District (CBD, QLD, Australia) run using OKTAL SCANeR™ software (OKTAL, France). Participants were required to react to a pedestrian crossing the road, on the first lap at an unsignalled crossing and on the second lap at a signalled crossing. Parked cars and pedestrians were placed on the side of the road to enhance the validity of the scenario. No other moving vehicles were included to ensure consistency of the scenario between participants and to assess free speed unimpeded by other vehicles. Participants were given a secondary signal detection task, to find a particular restaurant and press the horn when the restaurant sign was first seen, amidst six distracter hotel and bar signs. This task was incorporated to increase the ecological validity of the CBD simulation paradigm, increase the cognitive load of the driving simulation task and to assess signal detection performance within this scenario.

Executive function data included in this report are n-1 and n-2 back task accuracy (proportion of identical matched digit trials correctly identified as such, at 1- and 2-digits back respectively) as an index of working memory ability (n-1 is assumed to be relatively easy for most individuals, whereas n-2 is moderately challenging); GoStop stop inhibition (proportion of 'Stop' trials correctly inhibited for each of the four levels of Stimulus Onset Asynchrony (SOA) – 50 ms, 150 ms, 250 ms and 350 ms) as an index of response inhibition; and response times to complete the Trail Making Test Part A (basic level; general) and Part B (more challenging; set-shifting ability). For the simulator, relevant data included speed indices (mean speed, SD speed, maximum speed and proportion of the route spent above the posted speed limit), lane control (SD lateral position shift) and performance on the signal detection task (distance from the target sign when the horn was first pressed, for each of the two laps).

Results and Discussion

Given the modest sample size and preliminary nature of the data, correlations were conducted between executive function performance measures and driving simulator performance measures. These were analysed separately for each induction group (rest versus stress). Significance was assessed using $p < .05$. After removal of outliers, all variables met relevant assumptions with the exception of the lane control variable and proportion of time spent above the posted speed limit, which breached normality. A log10 transform successfully corrected these distributions. A manipulation check confirmed that post-induction VAS stress scores were greater in the stress induction group ($M = 35.63$, $SD = 19.99$) than the rest induction group ($M = 23.09$, $SD = 20.39$), $p < .05$. Selected correlations which exemplify the differences between rest and stress induction conditions are presented in Table 3.1. The full array of correlations is available on request from the first author.

Key Findings

Overall, few correlations reached statistical significance for the rest condition, although combined with trends ($r > .3$) the results suggest relationships between different aspects of young drivers' executive function and driving performance. Poorer performance on the TMT-A (longer completion times), a basic route tracing task, was associated with less variability in speed; while poorer performance on the TMT-B, a more challenging set-shifting task, was associated with less lane position variability. While both variables could be considered to reflect a safer style of driving, they could alternatively suggest that such poorer executive function is reflected in a reduced ability to adapt one's driving to the conditions,

Table 3.1 Selected correlations between executive function measures and driving simulator performance (*p* values in brackets)

	Rest condition								Stress condition							
	TMT-A	TMT-B	GS St50inhib	GS St150inhib	GS St250inhib	GS St350inhib	n-1 back acc	n-2 back acc	TMT-A	TMT-B	GS St50inhib	GS St150inhib	GS St250inhib	GS St350inhib	n-1 back acc	n-2 back acc
Mean speed	-.10 (.70)	.06 (.83)	.07 (.78)	-.01 (.96)	-.15 (.55)	-.47* (.05)	-.14 (.61)	-.23 (.35)	-.13 (.59)	-.29 (.21)	-.55** (.01)	-.21 (.38)	-.17 (.47)	-.08 (.74)	-.44 (.06)	.09 (.72)
SD speed	-.51* (.03)	.05 (.85)	.20 (.42)	.27 (.27)	.43 (.08)	.35 (.15)	.11 (.67)	-.33 (.17)	-.17 (.47)	-.04 (.85)	-.22 (.34)	-.18 (.44)	-.27 (.25)	-.09 (.72)	-.31 (.20)	.00 (.99)
Maximum speed	-.05 (.83)	-.09 (.73)	-.28 (.26)	-.20 (.42)	-.30 (.23)	-.19 (.46)	.09 (.74)	-.39 (.11)	.03 (.90)	.13 (.57)	-.40 (.08)	-.18 (.44)	.00 (.99)	-.29 (.21)	-.46* (.05)	-.13 (.60)
Time over speed limit	-.17 (.50)	.02 (.93)	-.30 (.23)	-.16 (.53)	-.30 (.23)	-.17 (.50)	.10 (.72)	-.31 (.21)	.07 (.78)	-.01 (.98)	-.42 (.06)	-.11 (.64)	.02 (.92)	-.26 (.27)	-.33 (.17)	.10 (.69)
SD lateral position	-.34 (.16)	-.59** (.01)	.20 (.43)	-.01 (.98)	.36 (.15)	.47* (.05)	.26 (.32)	-.08 (.76)	-.36 (.12)	-.08 (.75)	.03 (.89)	.25 (.28)	.19 (.43)	.16 (.51)	-.15 (.54)	-.02 (.93)
Signal detection lap 1	.26 (.31)	-.11 (.66)	.02 (.92)	-.40 (.10)	-.39 (.11)	-.38 (.12)	.27 (.32)	-.04 (.87)	.44* (.05)	-.02 (.92)	-.25 (.30)	-.16 (.50)	-.32 (.17)	-.09 (.71)	.07 (.78)	.03 (.89)
Signal detection lap 2	.22 (.40)	.01 (.98)	-.01 (.96)	-.50* (.04)	-.39 (.12)	-.27 (.29)	.46 (.09)	.05 (.85)	.46* (.04)	.26 (.28)	-.26 (.28)	-.14 (.54)	-.21 (.37)	.01 (.95)	-.14 (.58)	-.13 (.59)

Note: TMT = Trail Making Task Part A (-A) and B (-B) response times (ms); GS = GoStop task, St[50]Inhib = % inhibition at [50], [150], [250], and [350]ms SOAs; n-[1] back acc = accuracy for all n-[1] and n-[2] back trials, RT = response time to correct matching n-back trials. ** Correlation is significant at the 0.01 level (2-tailed). * Correlation is significant at the 0.05 level (2-tailed).

such as not reducing speed in the presence of pedestrian crossings. Neither of these relationships, however, was observed under stress induction.

Significant and trend correlations for the rest condition also revealed a consistent association between poorer stop inhibition across SOAs of 150 ms through to 350 ms (an indication of poorer response inhibition or impulse control) and greater distances at which the horn was pressed in response to identifying the target restaurant sign. While this could similarly appear to reflect better performance on the signal detection task, it may also indicate hypervigilance to the secondary task (possibly at the expense of the main driving task) or poorer impulse control which resulted in the horn being pressed earlier than the sign was actually identified by the participant. The first alternative is supported by the correlations observed in the rest group, albeit less consistent, between poorer stop inhibition at later SOAs (350 ms and 250 ms) and higher (but less variable) mean and maximum speeds, a greater proportion of the driving session spent above the speed limit and less lane position variability. Again though, these relationships disappeared under stress conditions. Finally, while not reaching statistical significance, trends ($r > .3$) in the rest condition data suggest a relationship between poorer working memory (lower n-2 back accuracy only) and greater speeds and speed variability; possibly reflecting a more erratic driving style due to a reduced capacity to hold and manipulate in memory the various requirements of the driving task (such as monitoring and adjusting speed, braking and steering; scanning, identifying and responding to hazards including pedestrians, traffic lights and the target sign).

Under stress, most of the observed relationships between executive function components and driving performance disappeared. However, significant correlations were observed at the more basic level of the n-back task (n-1, whereby the participant is required to remember and identify if the digit on the screen matches the previous, or '1-back', digit) and the GoStop stop inhibition task (50 ms SOA, at which it is typically easy to inhibit the prepotent response), with poorer performance on each associated with higher mean and maximum driving speeds. In contrast, poorer performance on the more basic TMT-A task was associated with greater distances at which the target restaurant sign was responded to via a horn press.

Conclusions

Overall, these preliminary findings highlight that driving performance is different for young adults under rest than under stress. Aspects of the city driving scenario employed in this study were designed by the authors to increase the cognitive load required of the driver, particularly the signal detection subtask which required participants to scan the environment and respond to the target signal (a restaurant sign, using the horn) amongst distractor signs, while simultaneously following the route and performing all relevant driving actions including steering, adjusting and maintaining appropriate speed, braking, turning and responding to pedestrian crossings and traffic lights. It was therefore expected, and supported by the results, that those young adults who had underdeveloped executive function would not cope as well as their more developed peers which would be reflected in poorer driving performance. However, while poorer executive function was typically associated with a poorer or riskier driving style in the simulator under rest conditions, such relationships were only seen at the extreme end of executive function deficits under stress. A potential explanation for this unexpected finding is the existence of a floor effect for those with more subtle baseline developmental deficits whereby the design was unable to detect even further stress-induced impairments added to their baseline impairment. Stress may have impaired the better executive functioning young drivers' performance to levels similar to those of 'rested' individuals with poorer executive function, in effect 'levelling' observed performance between the two groups under stress. In other words, a stressed young driver with 'normal' executive function may manifest the same risk on road as a rested young driver with poor or delayed executive function. While these preliminary results require replication in a larger sample, these findings highlight the influence of stress and of delayed or reduced executive function on young adults' driving performance and suggest these two factors should be addressed by the driver training and education field.

Acknowledgements

The authors gratefully acknowledge the assistance of Rachel Morton and Lauren Cunningham with data collection, Andrew Haines, Gregoire Larue and Husnain Malik with programming aspects of the CBD driving scenario, Luke Daly for his graphics design work on the CBD scenario, and Queensland University of Technology which provided the funding.

References

Australian Institute of Health and Welfare (2007). *Young Australians: their health and well-being 2007*, Cat No: PHE 87, AIHW, Canberra. Available from: http://www.aihw.gov.au/publications/aus/yathaw07/yathaw07.pdf

Barkley, R.A., Murphy, K.R., Dupaul, G.J., and Bush, T. (2002). Driving in young adults with attention deficit hyperactivity disorder: Knowledge, performance, adverse outcomes and the role of executive functioning. *Journal of the International Neuropsychological Society, 8*, 655–72.

Blakemore, S.J., and Choudhury, S. (2006). Development of the adolescent brain: Implications for executive function and social cognition. *Journal of Child Psychology and Psychiatry, 47*, 296–312.

Dougherty, D.M., Mathias, C.W., Marsh, D.M., and Jagar, A.A. (2005). Laboratory behavioral measures of impulsivity. *Behavior Research Methods, 37*, 82–90.

Fan, Y., Tang, Y., Lu, Q., Feng, S., Yu, O., Sui, D., Zhao, Q., Ma, Y., and Li, S. (2009). Dynamic changes in salivary cortisol and secretory immunoglobulin: A response to acute stress. *Stress and Health, 25*, 189–94.

Goldman-Rakic, P.S., Muly, E.C., and Williams, G.V. (2000). D1 receptors in prefrontal cells and circuits. *Brain Research Reviews, 31*, 295–301.

Knops, A., Nuerk, H-C., Fimm, B., Vohn, R., and Willmes, K. (2006). A special role for numbers in working memory? An fMRI study. *NeuroImage, 29*, 1–14.

Mitchell, R.L.C., and Phillips, L.H. (2007). The psychological, neurochemical and functional neuroanatomical mediators of the effects of positive and negative mood on executive functions. *Neuropsychologia, 45*, 617–29.

Morris, L.J., and Dawson, S.J. (2008). Relationships between age, executive function and driving behaviour. *Australasian Road Safety Research, Policing and Education Conference*, Centre for Automotive Safety Research, Adelaide. Available from: http://www.rsconference.com/pdf/RS080004.pdf

Pruessner, J.C., Champagne, F., Meaney, M.J., and Dagher, A. (2004). Dopamine release in response to psychological stress in humans and its relationship to early life maternal care: A positron emission tomography study using [11C] raclopride. *Journal of Neuroscience, 24*, 2825–31.

Sackler Institute for Developmental Psychobiology. (2007). *Face Go/No-Go Task.* New York: Cornell University. Available from: http://www.sacklerinstitute.org/cornell/assays_and_tools/

Spreen, O., and Strauss, E. (1998). *A Compendium of Neuropsychological Tests: Administration, Norms, and Commentary* (2nd Ed.). New York: Oxford University Press.

Smart, D., Vassallo, S., Sanson, A., Cockfield, S., Harris, A., Harrison, W., and McIntyre, A. (2005). *In the Driver's Seat: Understanding Young Adult's Driving Behaviour*, Research Report No: 12, Australian Institute of Family Studies, Melbourne. Available from: http://www.aifs.gov.au/institute/pubs/resreport12/aifsreport12.pdf

White, M.J. (2009). *Learning to Drive with Cognitive Impairment: the experience of young drivers and their parent supervisors*, Australasian Road Safety Research, Policing and Education Conference, Sydney. Available from: http://www.rsconference.com/pdf/RS094077.pdf

White, M.J., Lawford, B.R., Morris, C.P., and Young, R.McD. (2009). Interaction between DRD2 C957T polymorphism and an acute psychosocial stressor on reward-related behavioral impulsivity. *Behavior Genetics, 39*, 285 – 295.

White, M.J., Morris, P., Lawford, B.R., and Young, R.McD. (2008). Behavioral phenotypes of impulsivity related to the ANKK1 gene are independent of an acute stressor. *Behavioral and Brain Functions, 4*, 54.

Wickens, C.M., Toplak, M.E., and Wiesenthal, D.L. (2008). Cognitive failures as predictors of driving errors, lapses, and violations. *Accident Analysis and Prevention, 40*, 1223–33.

Chapter 4

Effects of Sadness on Drivers' Behaviour: An Empirical Study Using Emotional Induction and a Driving Simulator

Christelle Pêcher, Céline Lemercier and Jean-Marie Cellier
CLLE-LTC, University of Toulouse, France

Introduction

Although traffic research offers considerable scope for studying the role of emotions in driving (for a review, see Pêcher et al., 2011), the role of sadness in drivers' behaviour has rarely been investigated. By definition, sadness is a reaction to a loss or to an unsatisfied need (Lazarus, 1991; Oatley and Johnson-Laird, 1987). The events that trigger this emotional state are therefore not related to the driving situation but are dependent rather on drivers' attributes and history. From a methodological perspective, this indirect relationship between sadness and driving is a major barrier to the development of empirical work. Nevertheless, an experimental alternative to natural sadness may be used and this study is devoted precisely to understanding the effects of sadness on drivers' behaviour, using emotional induction and a simulated driving task.

In correlational and epidemiological studies, sadness and depression have been found to be good predictors of dangerous driving and road accidents (Dula and Geller, 2003; Garrity and Demick, 2001; Hilakivi et al., 1989). In addition, the fact of being involved in a negative and dramatic stressful event, such as the death of a relative or a divorce, has been associated with a higher risk of car crash (Lagarde et al., 2004; McMurray, 1970; Selzer and Vinokur, 1974). The evidence is that sadness, especially when due to a dramatic event, affects driving behaviour.

Indeed, in two empirical studies using a driving simulator, sadness and depression were found to have deleterious effects on driving controls (Bulmash et al., 2006). As an example, Pêcher et al. (2009) demonstrated that the presentation of sad music during a simulated driving task induced sadness in drivers and led to a deterioration of driving controls on speed. This reaction was explained in terms of impaired attention due to the presence of sad thoughts which captured the drivers' attention and interfered with attentional processes (for example, Huffziger and Kuehner, 2009; Nolen-Hoeksema, 1991). Also, from a methodological perspective, music was used here as a way of inducing sadness while driving, which offers a promising alternative to natural sadness.

Emotional induction procedures have been widely used in research on sadness and depression (for example, Clark, 1983; Scherrer and Dobson, 2009; for a review, see Gerrards-Hesse et al., 1994; Gilet 2008). According to Martin (1990), procedures that require mental imagery are among the most efficient. This category notably includes autobiographical recall, such as retrieving sad personal events and feeling related emotions (Brewer et al., 1980) and another procedure that requires the participant to imagine himself in a series of fictive dramatic situations and to focus on his emotions. These situations are presented in a form of sentences or 'vignettes' with sad music played in the background to promote a dramatic environment (Mayer et al., 1995). Both are easy to administer but the intrusive nature of autobiographical recall should be emphasized as well as the difficulty of imagining unreal situations in Mayer et al.'s procedure.

With regard to the above, the principal aim of this study was to measure the relative efficiency of these two procedures for inducing sadness in drivers. As music is known to provoke emotions (Krumhansl, 2002), we played music during the driving task to maintain, and indeed to increase, the effects of sadness induction (Mayer et al., 1995; Sutherland et al., 1982). The second aim of this experiment was to better understand the impact of the induced sadness on drivers' attentional behaviour, using a driving simulator.

Methods

Participants

Twenty-two French qualified drivers who were aged from 21 to 35 years old (12 females and 10 males; M = 26.6; SD = 3.5) from the University of Toulouse voluntarily participated in this study. They had all held a valid driving licence for at least four years (M = 7.4; SD = 2.6) and drove a minimum of 10,000 kilometres per year (M = 14,340 km; SD = 8,890 km). All reported having normal or corrected-to-normal vision. They were randomly assigned to one of the three induction groups: neutral induction (control group, n = 8), sadness induction with autobiographical recall (n = 7) or sadness induction with the vignettes-and-music procedure (n = 7).

Emotional Induction and Assessment Changes

For each induction, an identical soundtrack of complex classical music was played which consisted of a series of eight sad one minute extracts (pre-tested using the Self-Assessment Manikin, SAM; Bradley and Lang, 1994). During induction, participants had to focus on the music and perform an additional task based on it. Furthermore, instructions were manipulated to provoke either a neutral state or sadness.

- *Neutral induction*: Participants had to focus exclusively on the complex orchestration and define both the number and the type of instruments for each one of the eight extracts (inhibition of the emotional aspect of music).
- *Sadness induction with autobiographical recall*: As for the neutral induction, the eight one-minute musical extracts were played. For each extract, participants had to focus on its emotional nature and to mentally recall sad personal events (Brewer et al., 1980).
- *Sadness induction with the combined vignettes-and-music procedure*: This consisted of the simultaneous presentation of the eight sad musical extracts and a series of eight vignettes presented on a computer screen for one minute each (Mayer et al., 1995). Each vignette described a dramatic situation (for example, 'you are told by a young relative that he has cancer and only has six months to live'). Participants had to imagine themselves in the situation described, with the emotions commonly felt in such a situation.
- *Assessment of the emotional state*: The Positive Affect and Negative Affect Schedule – Expanded Form (PANAS-X; Watson and Clark, 1994) consists of 60 items measuring emotions and feelings, which are rated on a 5-point Likert scale which ranges from 1 'Very slightly or not at all' to 5 'Extremely', depending on the extent to which participants felt these emotions at the time. Although the PANAS-X distinguishes between 11 specific emotional states, we were only interested in sadness (5 items: *sad, blue, downhearted, lonely and alone*) and joviality (8 items: *happy, cheerful, delighted, joyful, enthusiastic, lively, peppy and excited*). For each specific state, scores were averaged across items. Participants completed the PANAS-X three times: before the induction, just after the induction and after the driving session. Additionally, participants were interviewed post-experiment to provide supplementary qualitative data.

The Driving Environment

The fixed-base driving simulator was located in the University of Toulouse II and consisted of a complete automobile (Renault 19) with an automatic transmission. It provided a 180° field of view. Both the training and the experimental circuits were collapsed-loop highways with two-lane traffic in each direction. The two scenarios used dry, daytime driving conditions with good visibility. The participant was the only driver on the road.

Participants were trained for approximately 15 minutes and were asked to respect French driving rules as they had to drive at 110–30 km/h, with a correct lane position. The experimental driving sessions lasted 15 minutes and 30 seconds with identical instructions. Throughout the sessions, musical extracts were played in the background. The series consisted of an alternation of five neutral extracts and five sad extracts of one and a half minutes each (also pre-tested with the SAM). Statistical analyses on mean speed and lateral position (standard deviation) were calculated for the first 30 seconds of each musical extract.

Advances in Traffic Psychology

Procedure

The experiment was run with each individual separately. On arrival, participants were randomly assigned to one of the three induction groups. They completed the first PANAS-X questionnaire and were trained for 15 minutes on the driving simulator. After that, they were induced and completed the second PANAS-X. They then performed the experimental driving session for 15 minutes and 30 seconds while sad and neutral music extracts were played in the background. After drivers had completed the last PANAS-X, they were interviewed and finally thanked for their participation.

Results

Assessment of the Emotional State: Scores of Sadness and Joviality on the PANAS-X

A repeated-measures analysis of variance (ANOVA) was performed on scores of sadness and joviality, including the within-subjects factor Stage of the experiment (before induction, just after induction and at the end of the experiment) and the between-subjects factor Induction group (neutral induction, sadness induction with autobiographical recall and induction with the combined vignettes-and-music procedure). Then, post-hoc analyses using Student t-tests for repeated measures were performed to determine changes of scores to the PANAS-X throughout the stages of experiment, for each induction group.

- *Sadness*: There was no effect on the Induction group ($F(2,19) = 1.89$; ns), but there was a main effect for the stage of the experiment ($F(2,38) = 7.65$; $p = .002$; $\eta^2_p = 0.28$), with an increase in sadness just after induction ($tri_1 = 1.20$; $tri_2 = 1.61$; $tri_3 = 1.20$). An interaction was also observed ($F(4,38) = 3.47$; $p = .01$; $\eta^2_p = 0.26$). Post-hoc analyses revealed no significant changes of sadness scores throughout the experiment for the control group ($t < 1$; ns) and for those who received sadness induction with autobiographical recall ($t < 1$; ns). A significant increase in sadness was observed just after induction with the combined vignettes-and-music procedure ($tri_1 = 1.02$; $tri_2 = 2.08$; $t(7) = -2.68$; $p = .01$) that decreased progressively during the driving session ($tri_2 = 2.08$; $tri_3 = 1.25$; $t(6) = 1.94$; $p = .05$). Nevertheless, some trace of the emotion induced remained, as there was still a significant difference between the beginning and the end of the experiment ($t(6) = -1.80$; $p = .05$).
- *Joviality*: There was no effect of the Induction group ($F < 1$), but there was a main effect for the stage of the experiment ($F(2,38) = 10.05$; $p = .000$; $\eta^2_p = 0.34$), with an unstable decrease in joviality just after the induction ($jov_1 = 3.09$; $jov_2 = 2.57$; $jov_3 = 2.57$). Finally, an interaction between the two factors was observed ($F(4,38) = 2.94$; $p = .03$; $\eta^2_p = 0.23$). Complementary

analyses revealed no significant changes in sadness scores throughout the experiment for the control group (t < 1; ns). After sadness induction with autobiographical recall, there were no significant changes between before and immediately after the induction (t < 1; ns) and between immediately after the induction and the end of the experiment (t < 1; ns). Nevertheless, a general decrease of joviality from the beginning to the end of the experiment was noted (jov$_1$ = 3.16; jov$_2$ = 2.67; t(6) = 4.40; p = .002). Finally, for those subjected to sadness induction with the combined vignettes-and-music procedure, a significant decrease in joviality was observed following the induction (jov$_1$ = 2.94; jov$_2$ = 2; t(6) = 6.57; p = .000), although it increased slightly during the driving session (jov$_2$ = 2; jov$_3$ = 2.37; t(6) = -2.29; p = .03). Nevertheless, there remained a significant difference in joviality scores between the beginning and the end of the experiment (jov$_1$ = 2.94; jov$_3$ = 2.37; t(6) = 2.11; p = 0.03).

Driving Parameters

A repeated-measures analysis of variance (ANOVA) was performed on mean speed and lane position (standard deviation), including the within-subjects factor Type of music (neutral music, negative music) and the between-subjects factor Induction group (neutral induction, sadness induction with autobiographical recall and sadness induction with the combined vignettes-and-music procedure).

- *Mean speed*: Analysis indicated a main effect for the 'Induction group' (F(2,19) = 4.51; p = .02; η^2_p = 0.32). Indeed, participants subjected to sadness induction with the combined vignettes-and-music procedure drove faster than the two other groups (respectively, MeanSpeed$_{Neutral}$ = 105.77 km/h; MeanSpeed$_{Recall}$ = 108.22 km/h; MeanSpeed$_{I+M}$ = 117.02 km/h). There was also a main effect for Type of music (F(1,19) = 6.55; p = .01; η^2_p = 0.25), with an increase in mean speed when sad music extracts were played (111.17 km/h) compared to neutral music extracts (109.09 km/h). Finally, there was no interaction between the two factors (F(2,19) = 0.35; ns).
- *Lane position standard deviation*: There was no effect of the Induction group (F < 1). Interestingly, analysis revealed a significant main effect of the Type of music (F(1,19) = 4.09; p = 0.05; η^2_p = 0.17). Drivers deviated less from their lane position when sad music extracts were played (1.92) compared to the playing of neutral music extracts (2.01). Finally, there was no interaction between the two factors (F < 1).

Discussion and Conclusion

In this chapter, the effects of sadness on driving behaviour were tested using an innovative experimental protocol which combined induction procedures, music and a simulated driving task. Analyses based on scores from the PANAS-X confirmed the stability of the control group (neutral induction). Surprisingly, autobiographical recall did not induce sadness whereas the use of the combined vignettes-and-music procedure was associated with a huge increase in sadness and a decrease in joviality, which were maintained during the driving session. These results support the hypothesis that this last procedure was efficient at inducing sadness in drivers. The lack of effect from the autobiographical recall could be explained by '*the level of intrusion*' of this procedure that probably interfered with the retrieval of events and emotions.

With regard to driving performance, results showed that participants who received sadness induction via the combined vignettes-and-music procedure drove 10 km/h faster than the two other groups. Nevertheless, they did not wander within their lanes and there were no speed differentials. Although this group drove faster (though remaining within speed limits), the absence of differences on other driving parameters indicated that they tended to anticipate dangers by stabilizing their performance. This hypothesis is supported by the fact that following sadness induction, drivers were more likely to have negative thoughts, which tended to make them inattentive. As they said in post-experiment interviews, they voluntarily inhibited thinking to keep their attention on the road and constantly react in an adaptive manner (see Converse et al., 2008; Schwarz and Clore, 1996).

Another interesting point concerns the playing of music while driving. Analyses on scores from the PANAS-X showed that music helped in maintaining sadness following the sadness induction with the combined vignettes-and-music procedure (Mayer et al., 1995; Pêcher et al., 2009; Sutherland et al., 1982). Nevertheless, drivers perceived all extracts as sad, without any distinction between neutral and sad music extracts. Here, perception and judgement of music suffered from a negative bias or an affective congruence effect (Avramova and Staple, 2008). This phenomenon favoured the maintenance of the effects of previous induction in promoting a sad, calm and moody musical environment when driving.

In conclusion, the results first showed that only the combined vignettes-and-music procedure was effective at inducing sadness and this may be reinforced by playing music while driving. In addition, only mean speed was affected, demonstrating its sensitivity to sadness. This empirical work offers new perspectives for traffic and experimental researchers as the use of emotional inductions would probably help in better understanding the role of sadness and thinking on drivers' attentional behaviour. Also, it appears to be a useful way to extend research to other emotions that have not yet been studied, such as joy and fear.

References

Avramova, Y.R., and Stapel, D.A. (2008). Moods as spotlights: The influence of moods on accessibility effects. *Journal of Personality and Social Psychology, 95*, 542–54.

Bradley, M.M., and Lang, P.J. (1994). Measuring emotion: The self-assessment manikin and the semantic differential. *Journal of Behavioural and Experimental Psychology, 25*, 49–59.

Brewer, D., Doughtie, E.B., and Lubin, B. (1980). Induction of mood and mood shift. *Journal of Clinical Psychology, 36*, 215–26.

Bulmash, E.L., Moller, H.J., Kayumov, L., Shen, J., Wang, X., and Shapiro, C.M. (2006). Psychomotor disturbance in depression: Assessment using a driving simulator paradigm. *Journal of Affective Disorders, 93*, 213–8.

Clark, D.M. (1983). On the induction of depressed mood in the laboratory: The evaluation and comparison of the Velten and musical procedure. *Advances in Behavioural Research and Therapy, 5*, 27–49.

Converse, B.A., Lin, S., Keysar, B., and Epley, N. (2008). In the mood to get over yourself: Mood affects Theory-of-Mind use. *Emotion, 8*, 725–30.

Dula, C.A., and Geller, E. (2003). Risky, aggressive, or emotional driving: Addressing the need for consistent communication in research. *Journal of Safety Research, 34*, 559–66.

Garrity, R.D., and Demick, J. (2001). Relations among personality traits, mood states and driving behaviours. *Journal of Adult Development, 8*, 109–18.

Gerrards-Hesse, A., Spies, K., and Hesse, F. (1994). Experimental induction of emotional states and their effectiveness: A review. *British Journal of Psychology, 85*, 55–78.

Gilet, A-L. (2008). Procédures d'induction d'humeurs en laboratoire: Une revue critique. *L'Encéphale, 34*, 233–9.

Hilakivi, I., Veilahti, J., Asplund, P., Sinivuo, J., Laitinen, L., and Koskenvuo, K. (1989). A sixteen-factor personality test for predicting automobile driving accidents of young drivers. *Accident Analysis and Prevention, 21*, 413–8.

Huffziger, S., and Kuehner, C. (2009). Rumination, distraction, and mindful self-focus in depressed patients. *Behaviour Research and Therapy, 47*, 224–30.

Krumhansl, C.L. (2002). Music: A link between cognition and emotion. *Current Directions in Psychological Science, 11*, 45–9.

Lagarde, E., Chastang, J.F., Guéguen, A., Coeuret-Pellicer, M., Chiron, M., and Lafont, S. (2004). Emotional stress and traffic accidents: The impact of separation and divorce. *Epidemiology, 15*, 762–6.

Lazarus, R.S. (1991). *Emotion and Adaptation.* New York: Oxford University Press.

McMurray, L. (1970). Emotional stress and driving performance: The effect of divorce. *Behavioural Research on Highway Safety, 1*, 100–114.

Martin, M. (1990). On the induction of mood. *Clinical Psychology Review, 10*, 669–97.

Mayer, J.D., Allen, J.P., and Beauregard, K. (1995). Mood inductions for four specific moods: A procedure employing guided imagery vignettes with music. *Journal of Mental Imagery, 19*, 133–50.

Nolen-Hoeksema, S. (1991). Responses to depression and their effects on the duration of depressive episodes. *Journal of Abnormal Psychology, 100*, 569–82.

Oatley, K., and Johnson-Laird, P.N. (1987). Towards a cognitive theory of emotions. *Cognition and Emotion, 1*, 29–50.

Pêcher, C., Lemercier, C., and Cellier, J-M. (2011). The influence of emotions in driving. In D.A. Hennessy (Ed.). *Traffic psychology: An international perspective*. New York: Nova Science Publishers.

Pêcher, C., Lemercier, C., and Cellier, J.M. (2009). Emotions drive attention: Effects on driver's behaviour. *Safety Science, 47*, 1254–9.

Schwarz, N., and Clore, G.L. (1996). Feelings and phenomenal experiences. In E.T. Higgins and A.W. Kruglanski (Eds.). *Social Psychology: Handbook of Basic Principles*. New York: Guildford Press.

Scherrer, M.C., and Dobson, K.S. (2009). Predicting responsiveness to a depressive mood induction procedure. *Journal of Clinical Psychology, 65*, 20–35.

Selzer, M.L., and Vinokur, A. (1974). Life events, subjective stress, and traffic accidents. *American Journal of Psychiatry, 131*, 903–906.

Sutherland, G., Newman, B., and Rachman, S. (1982). Experimental investigations of the relations between mood and intrusive, unwanted cognitions. *British Journal of Medical Psychology, 55*, 127–38.

Watson, D., and Clark, L.A. (1994). *The PANAS X: Manual for the Positive and Negative Affect Schedule- Expanded Form*. The University of Iowa: Unpublished manuscript.

PART II
Driver Distraction and Inattention

Chapter 5

A Roadside Survey of Driving Distractions in Austria

Mark Sullman
Department of Integrated Systems, Cranfield University, UK

Max Metzger
School of Psychology, University of Hertfordshire, UK

Introduction

Research has found that driver distraction is a contributing factor in at least 25 per cent of motor vehicle crashes (McEvoy et al., 2007; Stutts et al., 2001; Wang et al., 1996). This, however, could increase in line with the increased use of portable technologies, such as in-vehicle navigation systems and smart phones. Therefore, it is vital that researchers and road safety authorities have a clear understanding of the types of distractions drivers currently engage in, their prevalence and the types of drivers who are more likely to become distracted.

Driver distraction can be defined as any secondary activity that draws the driver's attention away from the main task of driving (Ranney, 1994). Although research investigating the effect of distractions on driving performance is relatively well developed, the research measuring exposure to driver distractions is relatively unadvanced (McEvoy and Stevenson, 2008). There are broadly four approaches to investigating driver exposure to distractions, with roadside observation being one of these approaches. Surprisingly there have been very few studies which have investigated driver distraction using roadside observation and most of those that do exist have solely focused on the prevalence of hand-held mobile phones (for example, Horberry et al., 2001; McEvoy et al., 2007; Taylor et al., 2007). Nevertheless, there has also been some research which has observed more general distractions using in-car observations, but these have mostly been conducted on lorry drivers in order to look at critical incidents and the distractions which lead up to these critical incidents (for example, Hanowski et al., 2005). There are currently only two peer reviewed studies which have looked more broadly at distraction amongst the general public.

In one of the two peer reviewed studies to look more broadly at the issue of driver distraction in car drivers, Stutts et al. (2005) used in-vehicle video cameras in 70 cars in the United States to investigate driver distractions. They found that the drivers in their study spent almost 40 per cent of their driving time engaged in some type of distracting activity. The most common distraction was conversing with passengers (15.3 per cent), followed by eating and drinking (4.6 per cent), smoking (1.55 per cent), manipulating controls (1.35 per cent) and using a mobile phone (1.30 per cent). As this research only included 70 drivers, it would be difficult to apply these results to the general public and it was not possible to test for any age or sex differences. This type of research is also very expensive, reliant on technology and is vulnerable to some form of experimenter effect. This latter issue was acknowledged by Stutts et al. when they reported that over 21 per cent of their drivers reported that their driving had been altered by having the equipment in their car.

One method of reducing the experimenter effect is to conduct the observations unobtrusively from outside of the vehicle. This was attempted by Johnson et al. (2004) who collected 40,000 high-quality still photographs of drivers passing through a New Jersey intersection. They found that less than five per cent of the photographs showed evidence of distraction, with the most commonly identified distraction being using a mobile phone, which was evident in one third of those drivers judged to be distracted. Smoking was the second most commonly observed distraction, with eating, drinking and interacting with a passenger accounting for most of the remainder. Surprisingly, Johnson et al. (2004) did not find any pattern for general distractions by age or sex, but they did find that younger drivers (< 45) were more likely to be observed using a hand-held mobile phone than older drivers.

As there is very little peer reviewed research investigating the issue of driver distraction using roadside observation and none currently from Austria, the present study set out to investigate the proportion of Austrian drivers who engage in an observable secondary task whilst driving. The research also set out to investigate the types of secondary tasks drivers engage in and whether there were any differences according to age, sex and time of day.

Method

Procedure

The data were collected via roadside observation using a clipboard, form and pen. The observer noted every vehicle that drove past and whether they were engaged in a secondary activity in addition to driving. Each of the observational sessions lasted for 60 minutes and a total of 20 observations took place in Salzburg (Austria). The age group of the driver, gender and time of day were also noted.

The observer was positioned so that the cars and the drivers were clearly visible, while at the same time being as unobtrusive as possible. In most cases the observer was not visible, in advance, to the motorist. In all occasions the observer monitored traffic which was coming towards them and on the same side of the road as they were. The observed secondary tasks had to fall within the previously established definitions to qualify as a distracter.

Definitions

> *Primary task only* – the motorist was engaged in driving only.
> *Mobile phone use* – the driver was holding a phone to their ear.
> *Texting/ Keying numbers* – the driver was clearly holding the mobile phone and pushing the keys in a manner to send a text message or dial numbers.
> *Drinking* – the driver was holding or drinking some form of beverage.
> *Eating* – the driver was holding or eating some form of food.
> *Smoking* – holding a cigarette and/or smoking it whilst driving a vehicle. This includes smoking, lighting and extinguishing a cigarette.
> *Other* – all other secondary tasks which did not fit into one of the previously mentioned categories, including: talking to a passenger, map reading, using a telecommunication device (other than a mobile phone) or reaching for something or adjusting controls (for example, heater, stereo, satellite navigation).

Locations

The observational sites were selected with the aid of an online random number generator. In the first step of this procedure a map of Salzburg was obtained and every street within the city limits was given a number. Following this 10 random numbers were generated using an online random number generator and these were matched to the corresponding road on the map, where the observations would take place.

Timing

All 10 sites were observed twice, once from 7 a.m. to 8 a.m. and secondly from 2 p.m. to 3 p.m. All observations were undertaken on a Monday or Friday. As the observations took place in the Austrian summer, the daylight and weather conditions allowed a clear view of the drivers on all occasions.

Results

In total 10,766 drivers were observed, with 8.7 per cent of these observed to be undertaking a secondary activity whilst driving. The most frequently observed secondary task was using a mobile phone, which was observed in 4.8 per cent of the cases (4.3 per cent talking and 0.5 per cent keying numbers or texting). The 'other' category contained the second largest proportion of distractions 1.7 per cent, followed by smoking at 1.4 per cent. Eating and drinking together accounted for the remaining 0.8 per cent.

Table 5.1 shows that the majority of both genders were engaged in the primary task (driving) only and that there were less females (3.7 per cent) than males (5.1 per cent) who were engaged in a secondary activity. A Chi-squared test revealed that these differences were not statistically significant χ^2 (1, 10765) = 1.3, p > 0.05.

A more detailed breakdown revealed that the most frequently occurring secondary task was the use of a mobile phone for both sexes (Table 5.2). In order to reach the minimum required cell count for all following analyses, the two mobile phone secondary tasks (phoning and texting) were combined, and the eating and drinking categories were combined. The second most frequently observed secondary activity for both genders was the 'Other' category, with 1.8 per cent for males and 1.7 per cent for females. The third most common distraction for both genders was smoking (both 1.4 per cent). As would be expected, these differences were not significant (χ^2 (3, 941) = 2.89, p > 0.05).

Table 5.3 shows the number of drivers who were observed driving only, versus those observed engaging in a secondary activity, by age group. There were statistically significant differences by age (χ^2 (2, 10766) = 55.90, $p < 0.001$). The standardized residuals in the distracted column are greater than +2.0 for both of the younger age groups, indicating that these two age groups were observed engaged in a secondary task more often than would be expected, if age were not related to

Table 5.1   Driving only versus involvement in a secondary task

Gender		Driving	Distracted	Total
Male	n (%)	5,510 (51.2)	546 (5.1)	6,056
	Std residual	-0.2	+0.7	
Female	n (%)	4,315 (40.1)	395 (3.7)	4,710
	Std residual	+0.3	-0.8	
Total		9,825	941	10,766

Table 5.2   Secondary activity while driving, by gender

		Secondary Task				Total
		Phone	Eat/Drink	Smoke	Others	
Male	n (%)	308(5.1)	41(0.7)	85(1.4)	112(1.8)	546
	Std residual	+0.5	-1.0	-0.3	+0.1	
Female	n (%)	209(4.4)	41(0.9)	66(1.4)	79(1.7)	395
	Std residual	-0.5	+1.1	+0.3	-0.1	
Total		527	82	151	191	941

distracted driving. In contrast, the standardized residual for the older age group (> 50) was -6.3 in the distracted column, indicating that older drivers were observed much less frequently engaged in a secondary task while driving.

Table 5.4 shows the type of secondary activity by age group. The observed secondary activities in the older drivers (> 50 years old) were slightly lower in most cases, except for smoking, which was more common in older drivers. In fact that standardized residual was +1.7, which is almost at the required +2.0 level. Surprisingly, although the proportion of older drivers using hand-held mobile phones was less than half that of the two younger age groups, this was not statistically significant, as shown by the low standardized residuals. The overall difference in distraction type, by age group, was not statistically significant (χ^2 (8, 10767) = 7.65, p > 0.05).

Table 5.3 **Driving only versus involvement in a secondary task, by age group**

Age group (years)		Driving	Distracted	Total
(< 30 Years)	n (%)	1,994(89.7)	230(10.3)	2,224
	Std residual	-0.8	+2.6	
(30–50 Years)	n (%)	5,646(90.4)	599(9.6)	6,245
	Std residual	-0.7	+2.3	
(> 50 Years)	n (%)	2,186(95.1)	112(4.9)	2,298
	Std residual	+1.9	-6.3	
Total		9,826	941	10,767

Table 5.4 **Secondary activity while driving, by age group**

Age group (years)		Secondary task				
		Mobile	Eat/Drink	Smoke	Other	Total
Young (< 30 years)	n (%)	122(5.5)	20(0.9)	31(1.4)	57(2.6)	230
	Std residual	-0.4	+0.0	-1.0	+1.5	
Middle (30–50 Years)	n (%)	337(5.4)	54(0.9)	95(1.5)	113(1.8)	599
	Std residual	+0.4	+0.2	-0.1	-0.8	
Older (> 50 Years)	n (%)	58(2.5)	8(0.3)	25(1.1)	21(0.9)	112
	Std residual	-0.5	-0.6	+1.7	-0.4	
Total		517	82	151	191	941

Table 5.5 shows that there appeared to be slightly more engagement in secondary activities in the morning than during the afternoon. However, a chi-square test revealed that the difference between the morning and afternoon observation times was not statistically significant (χ^2 (1, 10767) = 2.24, p = 0.134) and none of the standardized residuals were greater than 2.0.

Table 5.6 shows that regardless of the time of day, using a mobile phone to talk was the most common distraction, followed by other and smoking. Interestingly there were statistically significant differences by time of day (χ^2 (3, 941) = 12.45, p < 0.01). Looking at the standardized residuals, eating and drinking was clearly lower in the morning (standardized residual -2.2) than would be expected if time of day was unrelated to distraction type.

Table 5.5　　Driving only versus involvement in a secondary task, by time of day

Time of day		Driving	Secondary	Total
Morning	n (%)	3,941(90.8)	401(9.2)	4342
	Std residual	-0.3	+1.1	
Afternoon	n (%)	5,885(91.6)	540(8.4)	6425
	Std residual	+0.3	-0.9	
Total		9,826	941	10,767

Table 5.6　　Secondary activity while driving, by time of day

Time of day		Secondary task				
		Phone	Eat/Drink	Smoke	Others	Total
Morning	n (%)	225(5.2)	22(0.5)	76(1.8)	78(1.8)	401
	Std residual	+0.3	-2.2	+1.5	-0.4	
Afternoon	n (%)	292(4.5)	60(0.9)	75(1.2)	113(1.8)	540
	Std residual	-0.3	+1.9	-1.3	+0.3	
Total		517	82	151	191	941

Discussion

The present study found that although the vast majority were engaged in the primary task (driving) only, a substantial proportion of drivers (8.7 per cent) were engaged in a secondary activity at the time of observation. This is considerably higher than the 5.5 per cent found in a similar study in the UK (Sullman, 2010).

In terms of engagement in specific secondary tasks, the present study found that using a mobile phone was the most commonly observed distraction (4.8 per cent), with speaking on a hand-held phone being much more common (4.3 per cent) than texting or keying in numbers (0.5 per cent). This is a worrying finding, as it shows that not only are a substantial proportion of the drivers ignoring the law, but also that they are putting themselves and others at an increased risk of crash involvement. This needs to be addressed by some means, such as education and enforcement.

The rate of mobile phone use found here (4.8 per cent) is considerably higher than the 1.5 per cent found in Perth (Horberry et al., 2001) and the 1.6 per cent found in Melbourne (Taylor et al., 2007). There could be a number of reasons for this, such as differences in methodology and the increasing level of mobile phone ownership. However, the proportion found in the present study is also substantially higher than the 2.6 per cent found in the UK using the same methodology and in the same year (Sullman, 2010). Perhaps Austrian drivers are less compliant to traffic laws or there is less enforcement of those laws in Austria. Future research should investigate whether this high rate applies to the rest of Austria and what the reasons are for this high rate.

The present research did not find any sex differences in overall driver distraction or type of distraction. This finding is partially in contrast to Sullman (2010), who found that in the UK male drivers were more likely to be distracted than female drivers, but that there were no differences by distraction type. This finding is, however, in agreement with the research by Johnson et al. (2004), who found no sex differences in distraction. However, the previous findings on mobile phone use, regarding sex differences, have been inconsistent. Several studies have reported that males use a hand-held mobile phone more often (for example, Horberry et al., 2001), while one study found females use a mobile phone more often (Crammer et al., 2007) and others have found no difference (for example, Taylor et al., 2007; Townsend, 2006). Possibly this inconsistency simply confirms that there is no real difference in mobile phone use between males and females and perhaps in general driver distraction.

Also in line with previous research (Johnson et al., 2004; Sullman, 2010) the present study found that older drivers were much less frequently observed engaged in secondary tasks. The higher rate amongst young drivers is particularly concerning as they already have a higher risk of crash involvement than older drivers. Therefore, targeting educational and enforcement resources specifically at younger drivers appears to be necessary. However, in contrast to the previous research (for example, Horberry et al., 2001; Pöysti et al., 2005; Sullman, 2010; Sullman and Baas, 2004; Taylor et al., 2007) the present study did not find a difference in hand-held mobile phone use by age. This was a surprising finding and requires further investigation.

The level of distracted drivers observed here is likely to be an underestimate of the true level of driver distraction, as only visible distractions could be observed. Therefore, any distractions which did not involve a visible action, such as talking on a mobile phone using a hands-free device, could not be recorded. Cognitive distractions, such as being deep in thought or reading roadside signs were also not able to be recorded using this methodology.

Another potential limitation of this study is that in some cases the driver may have noticed the observer and changed their driving behaviour. However, as the observer was as unobtrusive as possible and in most cases was not likely to have been noticed by the driver, this is unlikely to have greatly influenced the data and findings. It should also be noted that the estimation of age was based solely upon the drivers' physical appearance, making this judgement completely subjective and relatively prone to error. Therefore, the age groups presented here should be thought of as approximate only.

References

Cramer, S., Mayer, J., and Ryan, S. (2007). College students use cell phones while driving more frequently than found in government study. *Journal of American College Health, 56*(2), 181–184.

Hanowski, R.J., Perez, M.A., and Dingus, T.A. (2005). Driver distraction in long-haul truck drivers. *Transportation Research Part F, 8*, 441–58.

Horberry, T., Bubnich, C., Hartley, L., and Lamble, D. (2001). Drivers' use of hand-held mobile phones in Western Australia. *Transportation Research Part F, 4*, 213–18.

Johnson, M.B., Voas, R.B., Lacey, J.H., McKnight, A.S., and Lange, J.E. (2004). Living dangerously: Driver distraction at high speed. *Traffic Injury Prevention, 5*, 1–7.

McEvoy, S., and Stevenson, M. (2008). Measuring exposure to driver distraction. In M.A. Regan, J.D. Lee and K.L. Young (Eds.), *Driver Distraction: Theory, Effects and Mitigation* (pp. 73–83). Boca Raton, FL: CRC.

McEvoy, S., Stevenson, M., and Woodward, M. (2007). The contribution of passengers versus mobile phone use to motor vehicle crashes resulting in hospital attendance by the driver. *Accident Analysis and Prevention, 39*, 1170–76.

Pöysti, L., Rajalin, S., and Summala, H. (2005). Factors influencing the use of cellular (mobile) phone during driving and hazards while using it. *Accident Analysis and Prevention, 37*, 47–51.

Ranney, T.A. (1994). Models of driving behavior: A review of their evolution. *Accident Analysis and Prevention, 26*, 733–50.

Stutts, J.C., Feaganes, J.R., Reinfurt, D., Rodgman, E., Hamlett, C., Gish, K,. and Staplin, L. (2005). Driver's exposure to distractions in their natural driving environment. *Accident Analysis and Prevention, 37*, 1093–1101.

Stutts, J. C., Reinfurt, D. W., Staplin, L., and Rodgman, E. A. (2001). *The Role of Driver Distraction in Traffic Crashes*. A Report prepared for the AAA Foundation for Traffic Safety, Washington, DC.

Sullman, M.J.M., and Baas, P. (2004). Mobile phone use amongst New Zealand drivers. *Transportation Research Part F, 7*, 95–105.

Sullman, M.J.M. (2010). An observational survey of driving distractions in England. *Driving Behaviour and Training IV.* (107–116). London; Ashgate.

Taylor, D.M.D., MacBean, C.E., Das, A., and Rosli, R.M. (2007). Handheld mobile telephone use among Melbourne drivers. *The Medical Journal of Australia, 187*, 432–4.

Townsend, M. (2006). Motorists' use of hand held cell phones in New Zealand: An observational study. *Accident Analysis and Prevention, 38*, 748–750.

Wang, J.-S., Knipling, R.R., and Goodman, M.J. (1996). The role of driver inattention in crashes: New statistics from the 1995 Crashworthiness Data System. In *40th Annual Proceedings Association for the Advancement of Automotive Medicine.* (pp. 377–92). Des Plaines, IA: AAAM.

Chapter 6

Personality and Demographic Predictors of Aggressive and Distracted Driving

Harold Stanislaw
California State University, Stanislaus, USA

Introduction

Drivers may exhibit various behaviours that greatly increase their likelihood of crash involvement. One group of such behaviours is often referred to as *aggressive driving*. This term is neither precise nor consistently applied (Dula and Geller, 2003), but the label is popular and was added to the PsycINFO *Thesaurus* in 2004. Aggressive driving thus invokes Justice Stewart's famous remark about pornography (Gerwitz, 1996): It is difficult to define but easily recognized. Examples of driving behaviours that researchers routinely classify as aggressive include: tailgating, speeding, running red lights, and failing to yield to other vehicles.

Formal definitions of aggressive driving are problematic in part because they tend to invoke both the behaviours in question and the underlying attributions for those behaviours. Confounding cause and effect in this manner is anathema to the goals of science, and in any event is premature, given our current understanding of aggressive driving. For example, Tasca (2000, p.2) suggests that driving behaviours qualify as aggressive only if they are 'motivated by impatience, annoyance, hostility and/or an attempt to save time'. However, Tasca himself acknowledges that speeding – which he (and others) consider aggressive – may result from sensation seeking and thus violate his own definition of aggressive driving. Similarly, Ellison-Potter et al. (2001, p.432) label driving aggressive when it 'intentionally ... endangers others psychologically, physically, or both' but it is not clear that all drivers who tailgate or fail to yield even realize that their behaviours are dangerous, let alone intend their behaviours to have particular effects on others. James and Nahl (2000, p.5) define aggressive driving as 'driving under the influence of impaired emotions, resulting in behaviour that imposes one's own preferred level of risk on others'. This definition is overly general; driving under the influence of alcohol would seem to qualify, although it is rarely considered a form of aggressive driving. Furthermore, the 'impaired' proviso suggests that aggressive driving is atypical, which belies the finding that aggressive behaviours are normative for some drivers.

As these examples suggest, a precise definition of aggressive driving is likely to remain elusive until a better understanding exists of the factors that underlie the behaviours of interest, and until clarity emerges regarding which driving behaviours cluster together as 'aggressive' and which seem to reflect disparate behaviours that justify a separate label. The present study attempts to address these needs by identifying personality and demographic characteristics that predict aggressive driving. No formal definition of aggressive driving will be proffered; instead, aggressive driving will be defined operationally as a specific set of behaviours that some drivers may exhibit more or less often than others, and that researchers consider to be aggressive as demonstrated by their own operational definitions (for an extensive list of example driving behaviours that qualify as aggressive see Table 1 in Van Rooy et al., 2006). The focus will be on examining relationships between the driving behaviours in question, to distinguish those that co-vary and are predicted by a common set of variables (thus supporting the existence of a single underlying psychological construct) from those that have unique predictors and otherwise seem independent of one another (and thus may reflect disparate psychological constructs).

As one might expect, the literature suggests that having an aggressive personality is predictive of aggressive driving (for example, Smith et al., 2006; Zuckerman and Kuhlman, 2000). However, 'aggression' is a nebulous term in the personality literature; many theorists subscribe to the view that four separate traits underlie aggression: physical aggression, verbal aggression, anger, and hostility (Bryant and Smith, 2001; Buss and Perry, 1992). Of these, anger (or the more situation-specific 'driving anger') has probably received the most study from traffic safety researchers and has repeatedly been shown to predict aggressive driving (for a meta-analytic review see Nesbit et al., 2007). However, anger is typically correlated with physical aggression, which may be a better predictor of aggressive driving (Krahé, 2005; Van Rooy et al., 2006).

Males and females have comparable levels of anger in general (Buss and Perry, 1992; Condon et al., 2006; Harris, 1996), and driving anger in particular (Lonczak et al., 2007; Smith et al., 2006). This would seem to predict the absence of sex differences in aggressive driving; however, males tend to drive more aggressively than females (for example, Dula and Ballard, 2003; Ellison-Potter et al., 2001; Oltedal and Rundmo, 2006; Smith et al., 2006). The discrepancy may be resolved by assuming that the tendency to drive aggressively is determined more by physical aggression than by anger, as males exhibit higher levels of physical aggression than females (for example, Buss and Perry, 1992; Condon et al., 2006; Harris, 1996). Even so, it may be that other sex differences exist in aggressive driving beyond those that can be attributed to an aggressive personality.

Similarly, a tendency towards a less aggressive personality with increasing age (Smith et al., 2006) may at least partially account for the finding that adolescents tend to drive more aggressively than older drivers (for example, Krahé and Fenske, 2002; Tasca, 2000). It may also be that age differences in aggressive driving reflect the willingness of adolescents to take more risks and seek higher levels of sensation than older drivers (Arnett et al., 1997; Clark et al., 2005; Machin and Sankey, 2008). Theories that posit risk taking and sensation seeking as origins of aggressive driving are bolstered by the finding that aggressive driving is often comorbid with other risky lifestyle behaviours, such as smoking, drinking, and failing to use a seat belt (Beck et al., 2007; Bina et al., 2006; Bingham and Shope, 2004).

An unresolved issue is the relationship between aggressive driving behaviours and behaviours that can distract the driver, such as texting or using a cell phone while driving. These latter behaviours have recently become of great concern, given the near ubiquity of potentially distracting technologies in vehicles. Arguing on purely theoretical grounds, Dula and Geller (2003) suggest classifying distracted driving together with certain forms of aggressive driving, such as tailgating without intending to harm others, under the general label 'risk-taking behaviours'. James and Nahl (2000) go even further, stating that distracted driving behaviours *are* aggressive driving behaviours. The limited empirical data that are available suggest aggressive and distracted driving are separate but related constructs; the two forms of driving are moderately correlated and share predictors that only partially overlap (Bone and Mowen, 2006).

A better understanding of the relationship between aggressive and distracted driving may help elucidate the psychological underpinnings of driving behaviours that elevate the risk of crash involvement. However, much of the research on distracted driving has focused on clarifying its attendant risks, such as determining how driving while using a hands-free phone compares with driving while using a hand-held device (for example, Ishigami and Klein, 2009). Less work has been conducted on identifying the types of drivers and driving situations in which distracted driving is particularly common.

What little work has been performed to date focuses almost exclusively on cell phone use while driving. For example, Beck et al. (2006, 2007) found that this behaviour was correlated with aggressive driving, suggesting at least some degree of commonality between aggressive and distracted driving. Similarly, on-road cell phone use appears to decrease with increasing age (Brusque and Alauzet, 2008; McCartt et al., 2003), mirroring the age-related trends in aggressive driving. However, sex differences in cell phone use while driving may be minimal or marginally higher in females than in males (McCartt et al., 2003; Townsend, 2006). This trend contradicts the higher rates of aggressive driving among males than females. Considered together, the results suggest that distracted and aggressive driving may be related but separate constructs.

Distracted driving can result from more mundane activities than utilizing technology behind the wheel. Drivers were eating and smoking behind the wheel for decades before cell phones became popular, but even less is known about the relationship of these potentially distracting behaviours to aggressive driving. On the one hand, smokers have long been known to crash more often than non-smokers (for example, DiFranza et al., 1986); on the other hand, the literature is silent on the issue of how often smokers engage in distracting behaviours while driving, and whether those who do so more frequently also drive more aggressively. Lacking this information, it is just as tempting to attribute the higher crash rates of smokers to greater risk tolerance or sensation seeking as it is to blame the distracting behaviours that smoking engenders. Indeed, Lonczak et al. (2007) note that smoking can simultaneously be considered both a distracter and a predictor of risky driving.

In light of these gaps in the literature, the present study sought to examine the degree to which drivers engage in behaviours that distract them while they are driving, and to relate trends in these behaviours to aggressive driving. This was intended to help establish whether or not aggressive and distracted driving are reflective of a single underlying construct. A second overarching goal was to determine which personality and demographic variables are capable of predicting the tendency to drive aggressively and in a distracted manner. Additional information was sought regarding the relationship of aggressive and distracted driving to self-rated driving ability and the perceived likelihood of crash involvement. This information was expected to provide insight into the degree to which drivers recognize the risks inherent in their driving styles.

Methods

Participants

Students in an undergraduate psychology research methods course at a public university in California's Central Valley (a region encompassing rural, suburban, and low-density urban areas) were asked to administer surveys to three male and three female drivers each. The drivers were required to be at least 18 years old and received no incentives for completing the survey. Some of the drivers were related to the students while others were acquaintances, co-workers, or strangers.

In five instances, drivers openly shared their responses with other participants or otherwise engaged in activities that threatened the validity of the data; these drivers were excluded from the study. Three other drivers were excluded because they completed only one side of the double-sided survey, yielding a final sample of 137 male and 137 female drivers. The sample was approximately evenly divided between university students (n = 134) and non-students (n = 139); one driver declined to provide this information. Age ranged from 18 to 67 years (M = 29.15, SD = 12.42), while driving experience ranged from 0 to 51 years (M = 11.86, SD = 12.45).

Materials

The survey opened with three physical aggression items and three anger items drawn from the Buss and Perry (1992) Aggression Questionnaire (Bryant and Smith, 2001). These were intermixed with five items from the Adolescent Invulnerability Scale (Lapsley and Hill, 2010).

Fourteen novel items designed to assess aggressive driving followed. Most of these involved behaviours commonly associated with aggressive driving (for example, 'I honk my horn at other drivers who irritate me' and 'I don't pay much attention to the speed limit on the freeway'), but two items examined seat-belt use (for example, 'When I'm a passenger in a car, I don't bother wearing a seat-belt'), which is typically associated with risky driving in general rather than aggressive driving in particular. Five of the aggressive driving items were reverse phrased.

The aggressive driving items were followed by one item that assessed self-rated driving ability ('I drive better than the average driver') and one item that assessed the perceived likelihood of crash involvement ('I will probably be involved in an accident in the next 12 months').

Ten distracted driving items were presented next. These included items relating to non-technological distractions (for example, 'I smoke while I'm driving' and 'I turn my head to talk to passengers while I'm driving') and items relating to the use of electronic devices (for example, 'I change the disk in a CD player while I'm driving' and 'I use a hand-held phone while I'm driving').

All 37 items described above used a 6-point response scale with anchors of 'strongly disagree' (1) and 'strongly agree' (6). The distracted driving questions also provided a 'not applicable' (0) response option that could be used when the driver lacked the electronic device mentioned by the item.

The survey ended with a series of questions regarding demographic characteristics and driving history. One such question asked whether the driver had ever received a moving violation, while another asked whether the driver had ever been involved in an accident that was at least partially his or her fault.

Procedure

Drivers read and signed a consent form, then completed the survey individually or in small groups. To ensure anonymity, the students who administered the surveys placed the signed consent forms in one envelope, while the drivers themselves placed their completed surveys in a separate envelope.

For scoring purposes, all 'not applicable' responses were recoded as 'strongly disagree', since the behaviour in question (for example, using a hand-held phone) could not be performed by a driver who lacked the device required to exhibit the behaviour. Reverse-phrased items were reverse-scored as needed. Average scores for each measure were then calculated, with a higher score indicating more frequent display of the behaviour in question. Drivers were classified as non-smokers if they chose the 'not applicable' or 'strongly disagree' response to the smoking item, and were classified as smokers if they chose any other response.

Results

Factor analysis of the 17 aggressive driving items yielded three factors with eigenvalues greater than one. However, the scree plot clearly indicated there was only one factor, which accounted for 28.4 per cent of the variance. Similarly, a factor analysis of the 10 distracted driving items yielded two factors with eigenvalues greater than one, but the scree plot supported only one factor; this accounted for 34.7 per cent of the variance.

Internal reliability was adequate for all measures (Table 6.1). Including the two seat-belt items as aggressive driving measures appeared to be justified; adding them increased the alpha coefficient from .78 to .80. With one exception, there was no evidence to suggest that deleting an item would markedly increase a measure's internal reliability. The sole exception was 'I use a hands-free phone while I'm driving'. Deleting this item, but retaining the similar item 'I use a hand-held phone while I'm driving' increased the distracted driving alpha from .74 to .79. This probably occurred because a given driver would tend to have either a hands-free phone or a hand-held phone, not both (the two phone items had a correlation of $r = .22$); thus, using one device would predispose the driver toward avoiding the other device. To accommodate this problem, the two cell phone items were combined by finding their maximum. This resulted in an alpha of .78 for distracted driving.

Table 6.1   Internal reliability, results of t-tests comparing drivers with and without citations, and with and without at-fault crashes

Variable	Alpha	No citations (n = 107)	Citation (n = 166)	p-value	No at-fault crashes (n = 173)	An at-fault crash (n = 99)	p-value
Aggressive driving	.80	2.34 (0.67)	2.60 (0.79)	.005	2.38 (0.72)	2.72 (0.77)	< .001
Distracted driving	.78	3.18 (1.06)	3.56 (1.11)	.005	3.18 (1.04)	3.83 (1.11)	< .001
Physical aggression	.73	2.68 (1.40)	2.84 (1.54)	.378	2.67 (1.42)	2.98 (1.57)	.094
Anger	.72	2.73 (1.18)	2.71 (1.18)	.872	2.57 (1.19)	2.97 (1.13)	.006
Invulnerability	.65	2.42 (0.91)	2.60 (0.94)	.123	2.60 (0.98)	2.41 (0.84)	.108
Self-rated ability	—	4.26 (1.25)	4.44 (1.33)	.270	4.52 (1.29)	4.14 (1.25)	.019
Crash likelihood*	—	1.51 (0.93)	1.76 (1.12)	.053	1.47 (0.89)	1.99 (1.22)	< .001

* Two drivers did not respond to this item

The item most predictive of aggressive driving was 'I tailgate cars that won't get out of my way', which had an item-total correlation of .57. The highest distracted driving item-total correlation was .64, for 'I read or send text messages while I'm driving'. Both of these correlations are sufficiently high that the two items could probably be used by themselves to measure aggressive and distracted driving, should the need arise for such a brief assessment.

Evidence for the validity of the aggressive and distracted driving measures was provided by comparing drivers who had received a traffic citation to drivers who had no such history; the former had significantly higher aggressive and distracted driving scores than the latter (Table 6.1). Similarly, drivers who were at least partially at fault in an accident had significantly higher aggressive and distracted driving scores than drivers who had not caused an accident.

Smoking was more common among males (28.5 per cent) than females (19.1 per cent); thus, analyses of variance (ANOVAs) were used to control for sex when comparing smokers with non-smokers, and to control for smoking status in evaluating sex differences. Compared to females, males had higher physical aggression, invulnerability, aggressive driving, and self-rated driving ability scores (Table 6.2); no significant sex differences were found for anger, distracted driving, or perceived likelihood of crashing. Smokers had significantly higher scores, than non-smokers, on all variables except self-rated driving ability. The significant

Table 6.2   Means (and SDs) and ANOVA results for driver sex and smoking status

Variable	Female non-smokers (n = 109)	Male non-smokers (n = 98)	Female smokers (n = 26)	Male smokers (n = 39)	p for main effect of sex	p for main effect of smoking	p for inter-action
Aggressive driving	2.24 (0.66)	2.56 (0.68)	2.68 (0.67)	2.99 (0.98)	.003	< .001	.956
Distracted driving	3.18 (1.10)	3.21 (0.93)	4.08 (1.16)	4.10 (1.04)	.862	< .001	.997
Physical aggression	2.42 (1.38)	2.83 (1.45)	2.77 (1.40)	3.69 (1.52)	.001	.004	.216
Anger	2.58 (1.14)	2.59 (1.21)	3.10 (1.13)	3.13 (1.16)	.935	.002	.944
Invulner-ability	2.23 (0.80)	2.69 (0.90)	2.43 (0.73)	3.03 (1.16)	< .001	.036	.580
Self-rated ability	3.87 (1.25)	4.85 (1.21)	4.19 (0.94)	4.67 (1.36)	< .001	.695	.160
Crash likelihood*	1.63 (1.02)	1.50 (0.86)	2.00 (1.29)	1.92 (1.35)	.492	.010	.854

* Two drivers did not respond to this item

main effect for smoking status on distracted driving was not an artefact caused by including smoking information in the distracted driving score, as similar p-values to those shown in Table 6.2 were obtained when distracted driving was calculated without including smoking information. None of the interactions between sex and smoking status approached statistical significance.

Aggressive and distracted driving were highly correlated with each other (r = .54). Most personality and demographic variables had significant first-order correlations with both aggressive and distracted driving; however, the correlations were stronger with aggressive driving than with distracted driving (Table 6.3). Multiple regression analyses found that aggressive driving could be predicted with greater accuracy (R^2 = .51) than distracted driving (R^2 = .38). Physical aggression, anger, invulnerability, age, sex, and perceived crash likelihood were all significant predictors of aggressive driving after controlling for distracted driving, while age and smoking status were the only significant predictors of distracted driving after controlling for aggressive driving.

Smoking status may have predicted distracted driving only because calculation of the distracted driving score included the six-point smoking information variable. When distracted driving scores were recalculated excluding this variable, the six-point smoking variable had a first-order correlation of .26 with distracted driving and the binary (yes/no) smoking status variable had a first-order correlation of .20. Both correlations were significant (p ≤ .001), but lower than the first-order

Table 6.3   **First-order correlations of variables with aggressive and distracted driving, and beta values for predicting aggressive and distracted driving**

Variable	Correlation with aggressive driving	Beta for predicting aggressive driving	Correlation with distracted driving	Beta for predicting distracted driving
Aggressive driving	—	—	.54**	.45**
Distracted driving	.54**	.36**	—	—
Physical aggression	.42**	.13*	.27**	.09
Anger	.38**	.16**	.19**	-.10
Invulnerability	.25**	.13**	.08	-.07
Self-rated ability	.07	.01	.05	.06
Crash likelihood	.35**	.21**	.23**	.05
Age	-.34**	-.15**	-.33**	-.16**
Sex	.24**	.16**	.04	-.11
Smoker	.27**	-.02	.35**	.21**

* p < .05, ** p < .01

correlation of .35 between smoking status and the original distracted driving measure. In the multiple regression analyses, smoking had a beta of .10 (p = .055) when the six-point variable was used to predict the revised distracted driving measure and .06 (p = .281) when the binary variable was used. These values are markedly less than the beta of .21 (p < .001) that was obtained when the binary smoking variable was used to predict the original distracted driving measure. Thus, smoking has questionable value as a predictor of distracted driving; it is perhaps better regarded as an indicator of distracted driving.

It is noteworthy that 78.1 per cent of the drivers in the sample reported using a hand-held phone and/or texting while driving, as both of these behaviours are illegal in California. Drivers and passengers in vehicles in California are also required to wear seat-belts, but 25.9 per cent of respondents admitted to violating this law on at least some occasions. Violators differed from non-violators in both aggressive and distracted driving. Aggressive driving scores were significantly higher (p < .001) in drivers who used a hand-held phone or texted while driving (M = 2.65, SD = 0.74) than in drivers who did not (M = 1.97, SD = 0.57). Similarly, distracted driving scores were significantly higher (p = .001) in respondents who violated the seat-belt law (M = 4.00, SD = 1.10) than in drivers who did not (M = 3.46, SD = 1.16).

Discussion

A limitation of this study is the reliance upon self-report; it is possible that the observed trends for personality and demographic variables reflect nothing more than the effect of these variables on the willingness to disclose negative behaviours. However, this interpretation implies that similar trends should have been observed for both aggressive and distracted driving, which was not the case. Furthermore, over three quarters of the respondents admitted violating the hand-held phone and texting laws (which were highly publicized when they were introduced) and more than one quarter acknowledged violating California's seat-belt law (despite an ongoing 'Click It or Ticket' campaign). The higher aggressive and distracted driving scores for drivers with a citation or at-fault crash lend further credence to the validity of the data.

Aggressive and distracted driving shared several characteristics but diverged in other respects. Both behaviours decreased with age and were seen more often in smokers than in non-smokers. By contrast, males drove more aggressively than females, while no sex difference emerged for distracted driving. It is tempting to conclude from these findings that aggressive driving is more critical than distracted driving in affecting crash involvement, since age, smoking status, and sex differences in aggressive driving mirror those seen in crash involvement, while the lack of a sex difference in distracted driving is contrary to established crash trends (for example, Clarke et al., 2005; DiFranza et al., 1986; Lonczak et al., 2007). However, aggressive and distracted driving are highly correlated, so that differences in one behaviour tend to reflect in the other. Thus, while sex may not affect distracted driving directly, it may do so indirectly by impacting aggressive driving. Despite their intercorrelation, aggressive and distracted driving clearly reflect separate underlying psychological constructs, as Bone and Mowen (2006) also found. Aggressive driving was predicted by three personality traits (physical aggression, anger, and invulnerability), while none of these traits directly predicted distracted driving.

An intriguing question for future research is why aggressive and distracted driving are so highly correlated. Are they linked by a third variable, such as sensation seeking or risk taking? Arguing against this is the failure of smoking status (which may reflect both sensation seeking and risk taking) to predict distracted driving. However, smoking status was determined indirectly from the item 'I smoke while I'm driving'; a direct measure might have yielded different results, allowing the identification of smokers who refrain from the activity while driving. More relevant, perhaps, is the finding that violators of the seat-belt law (a measure of aggressive driving) had higher distracted driving scores, while violators of the hand-held phone and texting laws (which are distracted driving behaviours) had higher aggressive driving scores. Violating a law is a form of risk taking, so aggressive and distracted driving may be linked by the willingness or desire to take risks.

The findings for self-rated driving ability and perceived likelihood of crash involvement are particularly interesting. Highly aggressive drivers viewed themselves as more likely to crash, while there was no such correlation for distracted driving. Self-rated driving ability was uncorrelated with both aggressive and distracted driving, even though the ability ratings had relatively large standard deviations; the respondents did not all simply rate themselves good drivers. This suggests that drivers do not base perceptions of their abilities behind the wheel upon the frequency with which they drive in a risky manner. Drivers seem to have at least some awareness that aggressive driving can lead to an accident, but they do not regard distracting behaviours as dangerous. Collectively, these results are concerning, but they also raise the possibility of developing interventions that might reduce distracted driving by making drivers aware of the hazards associated with diverting attention from the road.

Acknowledgements

I thank my 2009 Fall semester PSYC 3000 students for assistance with data collection and entry.

References

Arnett, J., Offer, D., and Fine, M.A. (1997). Reckless driving in adolescence: 'State' and 'Trait' factors. *Accident Analysis and Prevention, 29*, 57–63.

Beck, K., Yan, F., and Wang, M.Q. (2007). Cell phone users, reported crash risk, unsafe driving behaviors and dispositions: A survey of motorists in Maryland. *Journal of Safety Research, 38*, 683–8.

Beck, K., Wang, M.Q., and Mitchell, M.M. (2006). Concerns, dispositions and behaviors of aggressive drivers: What do self-identified aggressive drivers believe about traffic safety? *Journal of Safety Research, 37*, 159–65.

Bina, M., Graziano, F., and Bonino, S. (2006). Risky driving and lifestyles in adolescence. *Accident Analysis and Prevention, 38*, 472–81.

Bingham, C., and Shope, J. (2004). Adolescent problem behavior and problem driving in young adulthood. *Journal of Adolescent Research, 19*, 205–23.

Bone, S., and Mowen, J. (2006). Identifying the traits of aggressive and distracted drivers: A hierarchical trait model approach. *Journal of Consumer Behaviour, 5*, 454–64.

Brusque, C., and Alauzet, A. (2008). Analysis of the individual factors affecting mobile phone use while driving in France: Socio-demographic characteristics, car and phone use in professional and private contexts. *Accident Analysis and Prevention, 40*, 35–44.

Bryant, F., and Smith, B. (2001). Refining the architecture of aggression: A measurement model for the Buss-Perry Aggression Questionnaire. *Journal of Research in Personality, 35*(2), 138–67.

Buss, A., and Perry, M. (1992). The Aggression Questionnaire. *Journal of Personality and Social Psychology, 63*, 452–9.

Clarke, D., Ward, P., and Truman, W. (2005). Voluntary risk taking and skill deficits in young driver accidents in the UK. *Accident Analysis and Prevention, 37*, 523–9.

Condon, L., Morales-Vives, F., Ferrando, P.J., and Vigil-Colet, A. (2006). Sex differences in the full and reduced versions of the aggression questionnaire: A question of differential item functioning? *European Journal of Psychological Assessment, 22*(2), 92–7.

DiFranza, J., Winters, T.H., Goldberg, R.J., Cirillo, L., and Biliouris, T. (1986). The relationship of smoking to motor vehicle accidents and traffic violations. *New York State Journal of Medicine, 86*, 464–7.

Dula, C., and Geller, E. (2003). Risky, aggressive, or emotional driving: Addressing the need for consistent communication in research. *Journal of Safety Research 34*, 559–66.

Dula, C., and Ballard, M. (2003). Development and evaluation of a measure of dangerous, aggressive, negative emotional and risky driving. *Journal of Applied Social Psychology, 33*, 263–82.

Ellison-Potter, P., Bell, P., and Deffenbacher, J. (2001). The effects of trait driving anger, anonymity, and aggressive stimuli on aggressive driving behavior. *Journal of Applied Social Psychology, 31*, 431–43.

Gerwitz, P. (1996). On 'I know it when I see it'. *The Yale Law Journal, 105*, 1023–47.

Harris, M. (1996). Aggressive experiences and aggressiveness: Relationship to ethnicity, gender, and age. *Journal of Applied Social Psychology, 26*, 843–70.

Ishigami, Y., and Klein, R. (2009). Is a hands-free phone safer than a handheld phone? *Journal of Safety Research, 40*, 157–64.

James, L., and Nahl, D. (2000). Aggressive driving is emotionally impaired driving. Available from http://www.aggressive.drivers.com/board/messages/25/47. html/

Krahé, B. (2005). Predictors of women's aggressive driving behavior. *Aggressive Behavior, 31*, 537–46.

Krahé, B., and Fenske, I. (2002). Predicting aggressive driving behavior: The role of macho personality, age, and power of car. *Aggressive Behavior, 28*, 21–9.

Lapsley, D., and Hill, P. (2010). Subjective invulnerability, optimism bias and adjustment in emerging adulthood. *Journal of Youth and Adolescence, 39*, 847–57.

Lonczak, H., Neighbors, C., and Donovan, D.M. (2007). Predicting risky and angry driving as a function of gender. *Accident Analysis and Prevention, 39*, 536–45.

Machin, M., and Sankey, K. (2008). Relationships between young drivers' personality characteristics, risk perceptions, and driving behaviour. *Accident Analysis and Prevention, 40*, 541–7.

McCartt, A., Braver, E.R., and Geary, L.L. (2003). Drivers' use of handheld cell phones before and after New York State's cell phone law. *Preventive Medicine: An International Journal Devoted to Practice and Theory, 36*, 629–35.

Nesbit, S., Conger, J.C., and Conger, A.J. (2007). A quantitative review of the relationship between anger and aggressive driving. *Aggression and Violent Behavior, 12*, 156–76.

Oltedal, S., and Rundmo, T. (2006). The effects of personality and gender on risky driving behaviour and accident involvement. *Safety Science, 44*, 621–8.

Smith, P., Waterman, M., and Ward, N. (2006). Driving aggression in forensic and non-forensic populations: Relationships to self-reported levels of aggression, anger and impulsivity. *British Journal of Psychology, 97*, 387–403.

Tasca, L. (2000). A review of the literature on aggressive driving research. Available from http://www.aggressive.drivers.com/board/messages/25/49.html/

Townsend, M. (2006). Motorists' use of hand held cell phones in New Zealand: An observational study. *Accident Analysis and Prevention, 38*, 748–50.

Van Rooy, D., Rotton, J., and Burns, T.M. (2006). Convergent, discriminant, and predictive validity of aggressive driving inventories: They drive as they live. *Aggressive Behavior, 32*, 89–98.

Zuckerman, M., and Kuhlman, D. (2000). Personality and risk-taking: Common biosocial factors. *Journal of Personality, 68*, 999–1029.

Chapter 7

Impact of Inattention Provoked by Sadness on Older Drivers' Behaviour

Céline Lemercier and Christelle Pêcher
CLLE-LTC, University of Toulouse, France

Introduction

Elderly drivers are over-represented in fatal traffic accidents (Evans, 1988). Attention failures with age are commonly invoked as a cause of such a fact (Staplin and Lyles, 1991). Indeed, elderly people suffer from a general slowing down of processing as well as a degradation of the inhibition mechanism, leading to an alteration of selective attention.

Attention deficits are a hold-all concept actually covering diverse driving situations that are worth being defined to better understand their specific effects. In a review of literature, Lemercier and Cellier (2008) investigated the specificity of various situations and subsequently defined three distinct attention deficits while driving (interference, distraction and inattention). Firstly, interference is linked to attention sharing when performing two or more tasks in the same period of time (for example, phoning while driving). Secondly, distraction is an automatic and brief orienting of the attention focus from the road to a sudden, salient, irrelevant and external stimulus (for example, on-road light flash). Lastly, inattention to driving is defined as a long-lasting endogenous orienting of attention on thoughts possibly leading to an alteration of the attention control of the primary task. Two classes of thoughts usually generate driver inattention (Nolen-Hoeksema et al., 2008): on the one hand, distractive thoughts refer to the processing of cognitive items (for example, modifying the date or time of a meeting) and, on the other hand, ruminative thoughts refer to the processing of emotional items associated with a specific emotional state (for example, when sad, people ruminate on the causes and the consequences of their sadness).

Sadness provoked by the experience of divorce, a separation, or the death of a close relative could therefore be responsible for a three times higher risk of becoming involved in a serious accident (Lagarde et al., 2004). According to Teasdale and Green (2004), sadness is an emotional state which involves a self-focus on thoughts, leading the subject to become inattentive to their current activity (such as driving). In a recent study, Pêcher et al. (2011) studied the impact of sadness on the attention of young drivers. They demonstrated that induced sadness paired with sad musical reinforcement (leading to ruminative thoughts) denigrates the orientation of attention function. Current research on inattention in driving has mostly focused on young adults. However, in the next 20 years, the number of elderly drivers (persons 70 and over) is predicted to massively increase in developed countries. It is therefore of great interest to study the impact of such an attention deficit on old drivers' behaviour.

A growing body of research suggests that the ability to regulate emotion increases with age (Charles and Carstensen, 2007). Compared to young adults, older people recover more quickly from negative emotional states and report superior emotional control (Samanez-Larkin et al., 2009; Tsai et al., 2000). According to the literature, one can expect that older adults will be less sensitive to sadness induction than their younger counterparts, implying a less substantial impairment of attention due to their sad emotional state. However, in psychopathological studies, sadness and depressed syndromes increase with age, as demonstrated by Clément and Léger (1996). Research has found the prevalence of depression to be 10–15 per cent among people older than 65 years old, while depression affects only 1 to 4 per cent of the general population (Clément and Léger, 1996).

Therefore, it seems important to evaluate the impact of inattention provoked by sadness on older drivers' attention. To test this hypothesis, 40 older drivers performed the Attentional Network Test (ANT; Fan et al., 2002) after being induced into a neutral or a sad emotional state.

Method

Participants

Forty older drivers (mean age = 70.3, range 65 to 78) participated in the experiment. All drove more than 5,000 km per year. They were divided into four groups according to the induction type (sad versus neutral induction) and the presence of musical reinforcement during the ANT task (sad music versus no music) (See Table 7.1).

Table 7.1   Distribution of participants among the four experimental conditions

	Induction groups			
	Control	Induction	Rumination	Induction + rumination
Participants	10	14	9	11

Materials

Emotional material: The induction of sadness was based on the Mayer et al. (1995) procedure, which combines the presentation of eight sad music excerpts with eight sad sentences (for example, 'the pet you are fond of is dead'). The neutral induction consisted of reading eight French proverbs (for example, 'tomorrow is another day'). Musical reinforcement of the sad emotion used a musical extract from *Alexander Nevsky* by Prokofiev (1938).

Assessment of the emotional state: The Brief Mood Introspection Scale (BMIS; Mayer and Gaschke, 1988) consists of a list of 16 adjectives characterizing 8 emotions: happiness, love, energy, well-being, fatigue, fear, sadness and anger. Each was rated on a four point Likert scale (1 = 'I absolutely do not feel' to 4 = 'I feel very strongly') the degree to which participants felt this way at the time. The BMIS was presented three times: before the induction, just after the induction and after the ANT.

Attentional Network Test: This spatial identification task consists of pressing the key corresponding to the location indicated by a central arrow on the target screen (for details, see Fan et al., 2002). A cue screen was presented 400 milliseconds before the target. Four types of cue (no cue, central cue, double cue and spatial cue) were used to permit the measurement of two of the three selective attention functions (alerting and orienting). By manipulating the orientation of the arrows on the target screen (making them congruent, incongruent or neutral to the central target arrow), the last function of executive control was tested.

Procedure

The experiment was run with each individual separately. Upon arrival, participants were randomly assigned to one of the four induction groups. The experiment was divided into five steps. In the first, the third and the fifth steps, participants were invited to complete the BMIS (Mayer and Gaschke, 1988). In the second step, participants were induced either into a neutral or sad emotional state. In the fourth step, they performed the ANT either with or without sad musical reinforcement.

Results

Assessment of the Emotional State

The scores of sadness and joviality, from the BMIS, are presented in Figure 7.1. A repeated-measures analysis of variance ANOVA was performed on the BMIS scores, including the within-subjects factor 'Stage of the experiment' (before induction, just after induction and at the end of the experiment) and the between-subjects factor 'Induction group' (control, induction, rumination, induction + rumination). The ANOVA only found a significant main effect for the stage of the experiment ($F(2, 80) = 6.48$; $p = .001$; $\eta^2_p = 0.14$). The BMIS score decreased significantly from the first presentation (before induction) to the last one (at the end of the experiment), showing a higher level of sadness. Post hoc tests revealed no significant changes in BMIS scores for the control group ($BMIS_1$. $BMIS_2$. t < 1/ $BMIS_2$. $BMIS_3$. t < 1). However, a significant decrease in BMIS score was observed after sad induction ($BMIS_1 = 11$, $BMIS_2 = 9.8$; $t(13) = 4.05$, $p < 0.01$/ $BMIS_2 = 9.8$, $BMIS_3 = 9.7$; $t < 1$) and after the sad induction was paired with sad musical reinforcement ($BMIS_1 = 10.8$, $BMIS_2 = 9.8$, $t(10) = 2.24$, $p < 0.05$/ $BMIS_2 = 9.8$, $BMIS_3 = 10.1$; $t(11) = 2.24$, $p = .05$), demonstrating an increase in sadness. Surprisingly, no significant change in BMIS score was observed for the rumination group ($BMIS_1$. $BMIS_2$. t < 1/ $BMIS_2$. $BMIS_3$. $t(8) = 1.51$, ns).

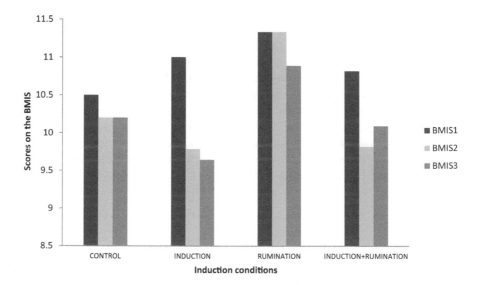

Figure 7.1   Scores on the BMIS at the three stages of the experiment as a function of the four induction conditions

ANT Performances

We carried out a 4 (Induction group: control, induction, rumination, induction + rumination) × 4 (Cue condition: no cue, centre cue, double cue, spatial cue) × 3 (Orientation type: neutral, congruent, incongruent) ANOVA of the RT data (see Table 7.2). There was firstly a significant main effect for the induction condition ($F(3, 40) = 11.09$; $p = .01$; $\eta^2_p = 0.45$), indicating a general increase in RT from control to other induction groups. There was also a significant main effect for cue condition ($F(3, 120) = 7.73$; $p = .01$; $\eta^2_p = 0.16$) and orientation type ($F(2, 80) = 85.05$; $p = .01$; $\eta^2_p = 0.68$). In addition, there was a significant interaction effect between the induction group and the orientation type ($F(6, 80) = 3.21$; $p = .01$; $\eta^2_p = 0.19$). Since our purpose was to evaluate the impact of sad induction and rumination on attention functions (alerting, orienting and conflict), we split the analyses according to the induction group. Post hoc analyses were then performed on each of the three attention functions for each induction group.

Control: A significant main effect of 39 ms (with a standard deviation of 16 ms) was found ($F(1, 9) = 18.35$; $p = .01$; $\eta^2_p = 0.67$). There was also an orienting effect of 47 ms (with a standard deviation of 21 ms) ($F(1, 9) = 7.54$; $p = .01$; $\eta^2_p = 0.45$). Finally, a conflict effect of 82 ms (with a standard deviation of 21 ms) was found ($F(1, 9) = 10.42$; $p = .01$; $\eta^2_p = 0.54$).

Induction: There was no alerting effect ($F < 1$). Nevertheless, there was a significant orienting effect of 49 ms (with a standard deviation of 14 ms) ($F(1, 13) = 6.69$; $p = .01$; $\eta^2_p = 0.34$) and a conflict effect of 149 ms (with a standard deviation of 31 ms) was also found ($F(1, 13) = 28.18$; $p = .01$; $\eta^2_p = 0.68$).

Rumination: Again there was no alerting effect ($F < 1$). Nevertheless there was a significant orienting effect of 74 ms (with a standard deviation of 16 ms) ($F(1, 8) = 14.63$; $p = .01$; $\eta^2_p = 0.65$) and a conflict effect of 161 ms (with a standard deviation of 7 ms) was found ($F(1, 8) = 18.28$; $p = .01$; $\eta^2_p = 0.70$).

Induction and rumination: A significant alerting effect of 51 ms (with a standard deviation of 2 ms) was found ($F(1, 10) = 8.04$; $p = .02$; $\eta^2_p = 0.45$). However, the orienting effect was no longer significant ($F < 1$). Finally, a conflict effect of 191 ms (with a standard deviation of 40 ms) was found ($F(1, 10) = 33.41$; $p = .01$; $\eta^2_p = 0.77$).

Table 7.2 shows the mean RTs in milliseconds for the three attention functions and error rates (in per cent) for each induction condition. The alerting effect was calculated by subtracting the mean RT of the double cue conditions from the mean RT of the no cue conditions. The orienting effect was calculated by subtracting the mean RT of the spatial cue conditions from the mean RT of the central cue conditions. Finally the conflict effect was calculated by subtracting the mean RT of all congruent cue conditions from the mean RT of all incongruent cue conditions.

Table 7.2   Mean RTs (msecs) for the three attention functions and error rates for each induction condition

	Induction conditions			
	Control	Induction	Rumination	Induction + rumination
Alerting				
Double	872	1,204	1,185	974
No cue	910	1,224	1,180	1,025
Difference	-38	-20	5	-51
Orienting				
Spatial	862	1,174	1,115	985
Central	909	1,224	1,189	1,040
Difference	-47	-50	-74	-55
Conflict				
Congruent	870	1,198	1,156	952
Incongruent	952	1,347	1,317	1,143
Difference	-82	-149	-161	-191
Error rates in % (SD)	3.02 (.67)	7.38 (.89)	13.31 (1.26)	14.52 (1.13)

Discussion and Conclusion

The present study evaluated the impact of inattention provoked by induced sadness on the three functions of attention for elderly drivers. Inattention is commonly defined as an attention deficit, linked to an endogenous orientation of the attention control on thoughts, leading to a degradation of the processing of the principal task. Every driver has had to face such a situation of inattention at some point (for example, trying not to forget to buy bread or to phone a friend). Referring to recent research about driving and emotions, inattention should also be linked to the sad emotional state associated with ruminations (Pêcher et al., 2010). In the present study, analysis of the BMIS data showed that sad induction led to an increase in sadness (as revealed by the lower BMIS scores in the induction, rumination, and induction and rumination conditions, compared to control). As a result of sad induction, older drivers were longer to perform the ANT (compared to controls), implying that they may have been preoccupied.

The ANT is used to measure the alerting, orienting and conflict functions of attention. Among the previous studies using the ANT, some of them studied normal young people, others ADHD patients, but none have focused on older drivers, even though these functions are critical while driving. The present study is then the first to evaluate those functions with elderly drivers. Here, aged drivers showed significant effects for alerting, orienting and conflict in the control condition, which is in line with the results of previous research (Fan et al., 2002).

As the control group is the reference group to which we compare the three other induction groups, we then evaluated the relative impact of the three conditions of sadness induction on the three functions of attention, as defined by Fan et al. (2002).

It is important to note that a very recent study by Pêcher et al. (2011) was conducted using the same protocol among young drivers. Their results demonstrated that inattention provoked by induced sadness, paired with sad musical extracts (promoting ruminative thoughts), led to a degradation of the orientation function of attention in the young drivers. The same results were found for the older drivers in the induction and rumination group. A possible practical issue is that spatial information on road signs, which are supposed to warn about the danger of a future spatially localized event on the road (such as lane change, or motorway exit), would not be processed and could have dramatic consequences when drivers are sad and ruminate on sad thoughts.

Finally, older drivers in the induction and rumination groups showed a reduction in the alerting effect. Alerting refers to the preparation of attention to the imminent appearance of the target. On the road, lots of road signs have such a warning function. They permit drivers to expect particular events and to anticipate the relevant driving action (for instance, slowing down the car before the next intersection, when a warning signal 'STOP in 500 metres' appears). According to our results, older drivers suffering from sadness or listening to sad music seem to be inattentionally blind to this type of warning information. This then represents a substantial concern for traffic and safety researchers.

As stated earlier, the ability to regulate emotion seems to improve across the adult life span (Charles and Carstensen, 2007). For instance, compared to young adults, older adults report superior emotional control (Tsai et al., 2000). In this study, however, analysis of the BMIS showed a degradation of the emotional state after sadness induction for older drivers, which has dramatic consequences on alerting and orienting. Referring to previous studies, we can conclude that sadness impacts attention processing not only for young drivers but also for older drivers in a more complex and diffused way. The present study is the first to evaluate the impact of a specific emotional state on the three critical functions of attention in driving amongst older people. Future research should also be conducted to investigate the impact of sadness (and other types of emotional states) on older drivers' attention in more ecologically valid conditions (in a real or simulated driving environment).

References

Charles, S.T., and Carstensen, L.L. (2007). Emotion regulation and aging. In J.J. Gross (Ed.), *Handbook of Emotion Regulation* (pp. 307–327). New York, NY: Guilford Press.

Clément, J.P., and Léger, J.M. (1996). Clinique et épidémiologie de la dépression du sujet âgé. In T. Lemperière (Ed.). *Les dépressions du sujet âgé.* Paris: Masson.

Evans, L. (1988). Older driver involvement in fatal and severe traffic crashes. *Journal of Gerontology, 43*(6), S186–93

Fan, J., McCandliss, B.D., Sommer, T., Raz, A., and Posner, M.I. (2002). Testing the efficiency and independence of attentional networks. *Journal of Cognitive Neuroscience, 14,* 340–47.

Lagarde, E., Chastang, J.F., Guéguen, A., Coeuret-Pellicer, M., Chiron, M., and Lafont, S. (2004). Emotional stress and traffic accidents: The impact of separation and divorce. *Epidemiology, 15,* 762–6.

Lemercier, C., and Cellier, J.M. (2008). Les défauts de l'attention en conduite automobile : Inattention, distraction, et interférence. *Le Travail Humain, 71,* 271–96.

Mayer, J.D., and Gaschke, Y.N. (1988). The experience and meta-experience of mood. *Journal of Personality and Social Psychology, 55,* 102–11.

Mayer, J.D., Allen, J.P., and Beauregard, K. (1995). Mood inductions for four specific moods: A procedure employing guided imagery vignettes with music. *Journal of Mental Imagery, 19,* 133–50.

Nolen-Hoeksema, S., Wisco, B.E., and Lyubomirsky, S. (2008). Rethinking rumination. *Perspective on Psychological Science, 3*(5), 400–24.

Pêcher, C., Lemercier, C., and Cellier, J.M. (2009). Emotions drive attention: Effects on driver's behaviour. *Safety Science, 47,* 1254–59.

Pêcher, C., Lemercier, C., and Cellier, J-M. (2010). The influence of emotions in driving. In D.A. Hennessy (Ed.). *Traffic Psychology: An International Perspective.* New York: Nova Science Publishers.

Pêcher, C., Quaireau, C., Lemercier, C., and Cellier, J-M. (2011). The effects of inattention on selective attention: How sadness and ruminations alter attention functions evaluated with the Attention Network Test. *European Review of Applied Psychology, 61,* 43–5.

Samanez-Larkin, G.R., Robertson, E.R., Mikels, J.A., Carstensen, L.L., and Gotlib, I.H. (2009). Selective attention to emotion in the aging brain. *Psychology and Aging, 24,* 519–29.

Staplin, L., and Lyles, R. W. (1991). Age differences in motion perception and specific traffic manoeuvre problems. *Transportation Research Record, 1325,* 23–33.

Teasdale, J.D., and Green, H.A.C. (2004). Ruminative self-focus and autobiographical memory. *Personality and Individual Differences, 36,* 1933–43.

Tsai, J.L., Levenson, R.W., and Carstensen, L.L. (2000). Autonomic, expressive and subjective responses to emotional films in younger and older adults of European American and Chinese descent. *Psychology and Aging, 15,* 684–93.

Chapter 8

Distracting Effects of Radio News and the Effects on Train Operator Performance

Masayoshi Shigemori and Ayanori Sato
Railway Technical Research Institute, Japan

Yusuke Shinpo and Nobuo Ohta
Gakushuin University, Japan

Introduction

On the 25 April 2005 in Japan, a train derailed and rammed into a nearby building killing 106 passengers, the driver and injuring 500 people. This incident is referred to as the Amagasaki rail crash. Investigators determined that the train driver's late braking at a curve was the primary cause of the accident. However, the reason for the late braking remains unclear (Aircraft and Railway Accidents Investigation Commission, 2007). The train driver's attention might not have been focused on operating a train, but on a radio conversation between the conductor and the operations commander, concerning his mistake in which he overran at a previous station. However, this factor remains speculative, since we do not know whether such listening could interfere with recognizing when to brake.

From an experimental psychological view, we can see the accident as a result of dual tasks with different modalities for the train driver; in other words, one task involves detecting visual braking points and braking, and the other task involves ignoring the conductor's radio conversation. We can refer to the former as a prospective memory task because the driver had to remember when and where to brake, while the latter can be categorized as a passive listening task, because the driver did not need to participate in the conversation.

This experimental situation is similar to driving while having a conversation on a cell phone. A large amount of research concerning the auditory effects on driving has been carried out and many studies have shown that hands-free cell phone conversations negatively affect car driving (see Collet et al., 2010a, 2010b). However, most of the studies, which have indicated an elevated risk from cell phone use, have required drivers to actively partake in the conversation (see Caird et al., 2008 for a review). Meanwhile the train driver only listened to the conversation and did not actively participate at the time of the accident. Active participation in conversation requires more cognitive skills than passively listening. In fact, many studies have shown passive listening to have little or no effect on car driving (Collet et al., 2009; McCarley et al., 2004; Strayer and Johnston, 2001).

Meanwhile a number of researchers have pointed out the importance of the conversation contents or subject matter. Simple conversations or subject matter of little interest to the driver/operator do not negatively affect driving/operating (Briem and Hedman, 1995). However, emotional subject matter has been found to affect driving more negatively than mundane conversations on a cellphone (Dula et al., 2011). In addition, researchers know that people are unable to disregard some types of auditory information, for example one's own name (Moray, 1959; Wood and Cowan, 1995).

In the Amagasaki accident, the driver had a strong interest in the conversation. Therefore, it is important to understand the effect of passively listening to material in which the participants have a strong interest. In this regard, it was only recently pointed out that important radio messages impacted the train driver's performance on a rapid visual serial presentation task and that the radio messages were memorised unconsciously (Ueda et al., 2010).

Previous studies related to car driving while using a cell phone have used various car-driving tasks, such as braking reaction time (Alm and Nilsson, 1995), signal detection (Hancock et al., 2003), lane-keeping (Brookhuis et al., 1991), or car following (Lamble et al., 1999). Prospective memory tasks, such as in the case of the Amagasaki rail crash, have not yet been investigated. Therefore, we investigated whether passively listening to material of strong interest to the driver would affect their prospective memory. It is important not only to understand the causes of the accident but also to highlight potential hazards in operating other equipment, such as automobiles, aeroplanes, or factory machinery. Furthermore, this research hopes to stimulate fundamental research into multimodal information processing and prospective memory.

Method

Participants

Forty-two undergraduates (18 males and 24 females, mean age = 21.12, SD = 1.43) voluntarily participated in the experiment. We allocated each of the 14 participants to one of three conditions of ignoring radio news in random order. The participants in the 'concerning' news condition engaged in a prospective memory task while listening to the radio news in which they indicated a high degree of interest/concern. Participants in the 'not concerning' news condition engaged in prospective memory tasks while listening to the radio news in which they indicated no strong interest/concern. The remaining participants in the 'white noise' condition engaged in prospective memory tasks while listening to 'white noise'. Therefore, this was a one factor between subjects experimental design which included three types of listening conditions.

Apparatus

Word stimuli for the prospective memory task were displayed at the centre of a colour monitor (Mitsubishi RD21EIII) controlled by an AV-tachistoscope system (Iwatsu ISEL IS702). Participants were required to press a particular key between one and five on the response box (Iwatsu ISEL IS-7211). A digital music player (ZEN X-Fi: ZN-XF16G-BK made by the Creative Technology Ltd) played the varying radio news via headphones.

Material

One hundred words were used in the prospective memory task. They were extracted from a word database developed by Amano and Kondo (2003). The words consisted of three to five Japanese katakana letters and highly familiar words. These words were ranked (according to their familiarity) at more than six on a seven point Likert scale (Amano and Kondo, 2003). Half of the words were presented to participants during the learning phase. During the test phase, those same words (old words) and the second half of the words (new words) were presented to participants to test their recognition. Also during the test phase, participants were instructed to press a certain key if they identified animal words, four of which were included in the new word set.

Radio news came from the internet based on 'concerning' or 'not concerning' news for each participant and was collected on the morning of the experiment or the night before the experiment. Participants' interests and concerns were previously identified during a preliminary survey.

Procedure

A few days before the experiment, we conducted a preliminary survey of the participants in order to identify their 'concerns' or things they were strongly interested in. During the survey they were asked to prioritize seven topics, which were as follows: politics, economics, show business, science, sports, crime and international affairs and to identify their favourite sport.

The experiment used a prospective memory task and an ignored radio news task. We used the event-based prospective memory task developed by Einstein and McDaniel (1990). It asked participants to engage in a word recognition memory task and if they identified a cue word in the word list of a recognition test, they were required to react by pressing a certain key.

At the start of the experimental session, we instructed participants on the procedure of the experiment, in particular the procedure for the prospective memory task. The learning phase followed this methodology: 50 words appeared one at a time at the centre of the display for three seconds at one-second intervals. The word list appeared twice and participants were asked to memorise them.

After the learning phase, participants were asked to undertake four calculations using double digit numbers as a filler task which induced participants to forget the memorised words. Then they were shown 100 words, one at a time, and asked whether they remembered those words from the learning phase. They were asked to press the 'old' key when they remembered the word or the 'new' key when they did not remember the word. The 'old' key was allocated to the first key and the 'new' key was allocated to the second key on the key unit. The words presented in the test included 50 old words and 50 new words. The 50 new words also included four animal words, as prospective memory cues. Participants were asked to press the fifth key on the key unit when they found cue words. Each session was conducted individually.

Results

In order to compare the mean number of correct responses in the prospective memory task, between the three passive listening conditions (Figure 8.1), an ANOVA was used. This indicated a trend towards a difference (F (2, 39) = 2.58, p < .1) and the result of the following post hoc comparisons showed a significant difference between scores from the 'concerning' news condition and those of the 'white noise' condition (t (39) = 2.27, p < .05). This means that the prospective memory score while listening to 'concerning' news was worse than for those who were listening to white noise. There were no differences between the other pairs of conditions.

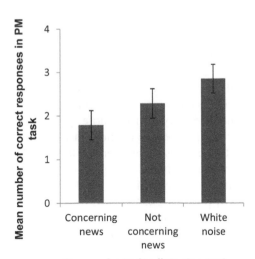

Figure 8.1   Mean number of correct responses by passive listening task

Participants also engaged in a recognition memory test within the prospective memory test. A mean correct recognition rate, by type of passive listening task, was calculated and analysed using an ANOVA (Figure 8.2). The results indicated a difference between conditions ($F_{(2, 39)} = 4.233$, $p < .05$), while the post hoc tests revealed that the recognition rate for the concerning news condition was significantly worse than that of the white noise condition ($t_{(39)} = 2.87$, $p < .05$). This means that the recognition test score while listening to 'concerning' news was worse than while listening to white noise. There were no differences between the other pairs of conditions.

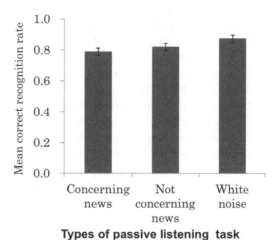

Types of passive listening task

Figure 8.2 **Mean correct recognition rate by type of passive listening (bars indicate standard error)**

Discussion

The prospective memory score was worse in the 'concerning' news condition than the white noise condition. This result suggests that even ignoring or passively listening to the radio can negatively affect prospective memory, if the content captures the participant's interest. Sometimes we are unable to ignore certain information, such as hearing our name mentioned during a cocktail party (Moray, 1959; Wood and Cowan, 1995). Reasoning suggests that information which has become habitually/automatically processed (automatic processing), can interfere with information processing with lower automaticity, in other words conscious control (controlled processing) (Schneider and Shiffrin, 1977; Shiffrin and Schneider, 1977). In particular, self-information strongly grabs our attention. Meanwhile, remembering to do things later stresses our attention resource, if they were not a part of routine work (McDaniels and Einstein, 2000); in other words it is rather controlled processing. Therefore, the concerning news, especially related to oneself, is processed automatically and interferes with typically controlled processing such as unfamiliar prospective memory.

In the Amagasaki accident, the train driver, who was a young novice, had to brake at a certain point. This represents a type of prospective memory task, which was not acquired by him sufficiently so that it enabled him to automatically recognize the point to begin braking. Some have speculated that his delayed braking was caused by interference from listening to the radio conversation concerning his error at the previous station. This experiment has demonstrated the plausibility of that speculation.

In addition, scores on the recognition task were also worse in the concerning news condition than that in the white noise condition. This result suggests that ignoring passively heard radio material may affect certain types of judgement related to memory. The recognition memory task asked participants to judge whether the presented words in the test phase were ones they had seen in the learning phase. This task included judgement and memory recall, which might be a type of controlled processing. Therefore, it is no wonder that listening to concerning news interferes with this type of judgement.

In the Amagasaki accident, it is not possible to know the driver's judgement. But in an ordinary driving situation we need various kinds of judgement, particularly when we drive on an unfamiliar route. For example, one has to judge the correct direction at junctions, lanes, or finding a parking space. According to our results, listening to the radio may negatively affect such decisions, if the driver finds the news concerning or interesting.

An abstracted prospective memory task was used in order to validate the presumed cause of the accident. Controversy exists when generalizing results from abstract laboratory work to practical problems. Therefore, further verification using a driving simulator or field study is required in order to confirm these results.

References

Aircraft and Railway Accidents Investigation Commission. (2007). *Tetsudo jiko tyosa hokokusyo: Nhisinihon ryokaku tetsudo kabushikigaisya Fukuchiyama-sen Tsukaguchi-eki - Amagasaki-eki kan ressha dassen jiko* [Railway accident analysis report: Derailment accident between Tsukaguchi station – Amagasaki station at West Japan Railway Company Fukuchiyama line]: Aircraft and Railway Accidents Investigation Commission.

Alm, H., and Nilsson, L. (1995). The effects of a mobile telephone task on driver behaviour in a car following situation. *Accident Analysis and Prevention*, *27*(5), 707–15.

Amano, S., and Kondo, T. (2003). *NTT Database Series Nihongo-no goitokusei* [Lexical properties of Japanese] CD-ROM (vol. 1). Tokyo: Sanseido.

Briem, V., and Hedman, L.R. (1995). Behavioural effects of mobile telephone use during simulated driving. *Ergonomics, 38*(12), 2536–62.

Brookhuis, K.A., de Vries, G., and de Waard, D. (1991). The effects of mobile telephoning on driving performance. *Accident Analysis and Prevention, 23*(4), 309–16.

Caird, J.K., Willness, C.R., Steel, P., and Scialfa, C. (2008). A meta-analysis of the effects of cell phones on driver performance. *Accident Analysis and Prevention, 40*(4), 1282–93.

Collet, C., Clarion, A., Morel, M., Chapon, A., and Petit, C. (2009). Physiological and behavioural changes associated to the management of secondary tasks while driving. *Applied Ergonomics, 40*(6), 1041–46.

Collet, C., Guillot, A., and Petit, C. (2010a). Phoning while driving I: A review of epidemiological, psychological, behavioural and physiological studies. *Ergonomics, 53*(5), 589–601.

Collet, C., Guillot, A., and Petit, C. (2010b). Phoning while driving II: A review of driving conditions influence. *Ergonomics, 53*(5), 602–16.

Dula, C.S., Martin, B.A., Fox, R.T., and Leonard, R.L. (2011). Differing types of cellular phone conversations and dangerous driving. *Accident Analysis and Prevention, 43*(1), 187–93.

Einstein, G.O., and McDaniel, M.A. (1990). Normal aging and prospective memory. *Journal of Experimental Psychology: Learning, Memory, and Cognition, 16*(4), 717–26.

Hancock, P.A., Lesch, M., and Simmons, L. (2003). The distraction effects of phone use during a crucial driving maneuver. *Accident Analysis and Prevention, 35*(4), 501–14.

Lamble, D., Kauranen, T., Laakso, M., and Summala, H. (1999). Cognitive load and detection thresholds in car following situations: Safety implications for using mobile (cellular) telephones while driving. *Accident Analysis and Prevention, 31*(6), 617–23.

McCarley, J.S., Vais, M.J., Pringle, H., Kramer, A.F., Irwin, D.E., and Strayer, D.L. (2004). Conversation Disrupts Change Detection in Complex Traffic Scenes. *Human Factors, 46*(3), 424–36.

McDaniels, M.A., and Einstein, G.O. (2000). Strategic and automatic processes in prospective memory retrieval: A multiprocess framework. *Applied Cognitive Psychology, 14*(SpecIssue), S127–S144.

Moray, N. (1959). Attention in dichotic listening: Affective cues and the influence of instructions. *The Quarterly Journal of Experimental Psychology, 11*, 56–60.

Schneider, W., and Shiffrin, R.M. (1977). Controlled and automatic human information processing I: Detection, search, and attention. *Psychological Review, 84*(1), 1–66.

Shiffrin, R.M., and Schneider, W. (1977). Controlled and automatic human information processing II: Perceptual learning, automatic attending and a general theory. *Psychological Review, 84*(2), 127–90.

Strayer, D.L., and Johnston, W.A. (2001). Driven to distraction: Dual-task studies of simulated driving and conversing on a cellular telephone. *Psychological Science, 12*(6), 462–6.

Ueda, M., Naito, H., and Usui, S. (2010). Musen renraku jushinji niokeru tetsudo untenshi no tyuui tokusei [Attention characteristics of train drivers when receiving train radio message]. *Japanese Journal of Ergonomics, 46*(1), 1–9.

Wood, N., and Cowan, N. (1995). The cocktail party phenomenon revisited: How frequent are attention shifts to one's name in an irrelevant auditory channel? *Journal of Experimental Psychology: Learning, Memory, and Cognition, 21*(1), 255–60.

PART III
Vulnerable Road Users

Chapter 9

Typical Human Errors in Traffic Accidents Involving Powered Two-Wheelers

Magali Jaffard and Pierre van Elslande

French Institute of Science and Technology for Transport, Development and Networks (IFSTTAR), France

The Problem of PTWs in the Traffic System

Powered two-wheelers (PTWs) are one group of road users who are particularly vulnerable to road accidents. In 2009 motorbikes accounted for 1.2 per cent of all motor vehicle traffic in France, while they accounted for 14 per cent of the vehicles involved in injury accidents and 23.5 per cent of motor vehicle fatalities (ONISR, 2009). In part, these figures result from the high level of PTWs 'fragility' due to the absence of protection provided by the vehicle body, as the slightest collision exposes the user to injury. However, these increased rates are also due to the greater involvement of PTWs in certain types of accidents. These two characteristics are testimony to the problem of the mismatch between PTWs and the traffic system. PTWs stand out in traffic because of their specific behaviours (for example, position on the carriageway, levels of acceleration, types of overtaking), particular dimensions, low representation in traffic, and a wide range of uses, in other words the practices and behaviours that depend on the type of PTW (for example, 50 cc, 125 cc or high-powered motorcycles). A better understanding of these problems requires an in-depth analysis of the difficulties met by PTW riders when interacting with the other roads users, notably on the basis of data collected on accident production mechanisms. This approach will also enable a better understanding of the difficulties that other road users encounter when meeting PTWs.

In fact, accidents shed light on the difficulties that drivers are unable to overcome in the driving situations they are confronted with. The detailed analysis of driver failures, of the contributing factors and of the accident circumstances, provide us with precious information on the difficulties facing PTWs and those confronted with them. The first section of this chapter is specifically dedicated to the results of such an analysis of the different types of operational problems in the interactions between PTWs and other users. In particular this clearly shows an over-representation of accidents involving the detection of PTW vehicles. The second section focuses on the problem of PTW conspicuity, as it is a problem commonly observed in such accidents.

Driving Failures, Factors and Contexts behind PTW Traffic Accidents

This first section is based on a detailed analysis of a sample of 384 PTW drivers involved in traffic accidents, a sample of 218 car drivers involved in a traffic accident with a PTW, and a control group of 1,174 car drivers involved in traffic accidents not including a PTW. This database underwent statistical weighting in order to provide a better picture of accident causation at the French national level. Accident data were collected by IFSTTAR multidisciplinary teams operating at the scene of the accidents. The investigations covered the three components of the road system: vehicles, drivers and infrastructure. Each accident investigated gave rise to a reconstruction in time and space of the events leading up to it. Then, a classification model of human functional failures (Van Elslande and Fouquet, 2007) was applied to the information collected by detailed interviews in order to define the breakdowns in the driver's functional chain (perception, diagnosis, prognosis, decision, execution and overall). The contextual explanatory elements behind these human failures (endogenous or exogenous) were also defined. This allowed a cognitive analysis of the 'human errors' to which the PTW riders and the other road users were exposed to. The accidents were analysed from the view point of the failures encountered by the PTW riders in their interactions with another user, and also from the view point of the specific failures of the other user (OU) involved in the accidents with the PTWs.

The samples studied included only 'interaction accidents'; in other words, accidents that involved an interaction problem between two (at least) road users. These interaction accidents were distinguished from cases involving a loss of control, which will not be dealt with here.

Problems involving the interaction with other road users were by far the most common problems encountered by PTWs. These types of accidents account for most of the PTW users included in this study (79 per cent). They involve motorcycles (53 per cent of them) as well as mopeds (47 per cent of them). These accidents usually correspond to complex phenomena. Understanding their mechanisms and the processes involved necessitates not only to take into consideration the PTW vehicle, the rider's capacities and the characteristics of the environment, but also to consider the other vehicle encountered, its driver's psychophysiological capacities and so on. Furthermore, the explanatory elements in the accident process are usually interlocking, and the error from one driver can influence an error by the other person involved. The present study was carried out in order to understand such phenomena.

Powered Two-Wheelers in Interaction with Others

The analysis of the interactions between powered two-wheelers and other actors in the driving system shows that certain failures are common to motorcycle and moped riders (notably failures relative to the prognosis made, such as how the situation will develop, or the poor understanding of the manoeuvre undertaken by the other user). However, other error profiles clearly stand out for motorcycle riders, moped riders and other road users in accidents where they interact. For example, while the most common failures among the control group of car drivers (not involved in a PTW accident) were detection failures (45 per cent); among PTW drivers these detection failures did not exceed 30 per cent. On the other hand, not predicting another user's manoeuvres was the most frequent functional failure among motorcycle riders (39.4 per cent of cases involving a motorcycle in interaction accidents), and making a poor decision to undertake a manoeuvre contrary to road safety rules was observed in one in five cases among moped drivers.

A variety of factors explain accidents involving PTWs, but there is little difference in the profiles of motorcycle and moped drivers. We observe that over-experience with the route, an overly rigid attachment to right-of-way status, speeds that are excessive for the situation and the atypical characteristics of the manoeuvres performed by others are at the heart of most accidents involving both moped and motorcycle riders.

A distinction can be made between motorcycle drivers and moped drivers in terms of their level of participation in accidents. For the most part, the former only contribute to the breakdown of the situation by not adopting appropriate adjustment strategies, while in more than half of all cases the latter are directly at the origin of the accident-causing disturbance. Thus, for motorcycle drivers, a relevant countermeasure would be to remind them that their priority status is not enough to protect them from an accident and train them to develop more compensatory strategies, taking into account their vulnerability. Among moped drivers, the accident mechanisms appear to be more linked to the characteristics of young drivers, such as risk-taking, trivialization of danger situations, and poor knowledge of safety rules. An appropriate countermeasure for them would be to increase awareness of the hazardous nature of this travel mode and behaviour. Such interventions will need to be adapted according to our understanding of how teenagers act in general.

Car Drivers Confronted with Powered Two-Wheelers

This last part of the analysis shows that accidents involving powered two-wheelers and their underlying processes are also specific for car drivers who are confronted with two-wheelers, as compared with other car drivers' accidents.

Dealing with functional failures, the results obtained show an obvious trend, in that there are far more detection problems among car drivers confronted with a PTW than for the control group of car drivers (59.9 per cent versus 45.0 per cent). Concerning driving tasks during which the accident occur, the results indicate that car drivers confronted with PTWs have greater difficulties when performing specific manoeuvres, such as overtaking or lane changing (35 per cent of the accidents) and when crossing an intersection (40 per cent), than drivers in the control sample (14 per cent and 37 per cent, respectively).

There are a variety of explanatory elements for failures among car drivers confronted with a motorcycle or moped which translate into the multiplicity of mechanisms underlying these accidents. Car drivers confronted with mopeds are more sensitive to traffic conditions (for example, atypical manoeuvres by the moped rider, elements in the environment hiding the presence of the moped) and to visibility problems. On the other hand, car drivers confronted with motorcycles appear to be more sensitive to their own attentional problems, such as over-experience with the route or with the manoeuvre to be performed or, on the contrary, their lack of knowledge regarding the location and situational constraints they come up against. These explanatory elements attest to the fact that the attentional resources given to the driving task are insufficient for correctly managing an interaction with a PTW.

Finally, an investigation of the level of involvement by drivers confronted with PTWs showed that their contribution to the accident was substantial when they interacted with a motorcycle, whereas for crashes involving a moped it was usually the moped driver who caused the breakdown in the situation.

But what stands out the most from such research on functional failures involved in accidents, is the crucial importance of the difficulties for car drivers to perceive PTWs. The second section of this paper specifically focuses on the problem of PTW conspicuity as a source of these difficulties.

The Problem of PTW Conspicuity and its Role in Traffic Accidents

As we have seen, cases involving a PTW and a third party show a detection problem for nearly 60 per cent of all car drivers involved in an accident with a PTW (while it is only 45 per cent for other car drivers' accidents). Along the same lines, the European MAIDS study (MAIDS, 2004) reported that 70 per cent of errors by car drivers, considered to be responsible for an accident involving a powered two-wheeler, were due to a failure to detect the PTW. To explain these figures, low detectability or low 'conspicuity' is a frequently mentioned accident-producing factor in the literature (Donne, 1990; Hurt et al., 1981; Preusser et al., 1995; Yuan, 2000). Accidents involving a perception problem, however, often concern a complex reality that needs to be further understood. Detection problems do not constitute a cause in and of themselves. They have their own causes, and the following analysis seeks to define them more precisely, with all their diversity.

Conspicuity is defined as the capacity that an object has, given its characteristics, to attract attention and to be easily located in its environment (Wulf et al., 1989). For road traffic, this corresponds to the capacity of the different users of the traffic space (for example, heavy goods vehicles, cars, two-wheelers, pedestrians) to be detected or perceived by the other users with whom they interact. This not only depends on their physical capacity to attract attention, but also on the observer's mental representations, knowledge, past experiences and objectives. Thus, perception is both a passive system that receives a quantity of incoming information ('bottom-up system') and an active system that conditions the way in which information is gathered within the environment ('top-down system'). The perceptual system therefore constitutes a filter for reality, which means that the perception one may have of a situation does not constitute a reproduction of reality, but rather a mental construction of reality among other possible constructions. It should be pointed out that, while use of the perceptual system leads to the detection of certain objects in the environment, detection of these objects alone does not ensure the correct processing of the visual information. Furthermore, it is common for PTWs to be detected on the road, but their distance and approach speeds may not be assessed properly by the observer. The problem of perceiving PTWs, due to their low conspicuity, does not just concern accident cases in which the PTW was not detected; poor PTW perception can have an impact on each stage of the information-processing chain (for example, situational understanding, decision-making).

Problems of perceiving a PTW on the road can be explained both by factors specific to the PTW (for example, behaviour, dimensions) and by a number of physical, cognitive and environmental factors which have no relation to the fact that it is a PTW (for example, the drivers psychophysiological state, their attention, the complexity of the infrastructure). Analysing the PTW conspicuity problem thus requires placing ourselves at the heart of the observer/PTW interaction.

Method for Analysing the Problem of Perception in Accidents

The following survey was based on an evaluation of the perceptual difficulties that drivers meet when they are confronted with a PTW. The survey makes use of an analysis performed on a sample of 127 accidents which involved an interaction between a PTW and another vehicle or pedestrian (accidents occurring between 1993 and 2009). The experts question in each accident case was not 'Is there a PTW conspicuity problem here?' but rather 'Is there a perception problem at some point in the accident sequence, regardless of whether the driver encountered a PTW or not?' Taken in this sense, the analysis allows us to go back to accidents involving a problem of PTW conspicuity and to base our claims on significant facts identified in the accident analysis. If a perception error was determined, the aim was to focus on the key factors thereof and to determine the specificity of these factors to PTWs.

The Problem of Perception and Interaction in the Error

Firstly, we established that perception problems were involved in the vast majority (78 per cent) of failures among road users confronted with a PTW. Obviously, these perception problems mainly led to detection failures (73 per cent of the failures). As can be seen in Table 9.1, the first detection failure dealt with 'Incomplete or hasty information gathering' (32 per cent), the second with 'Information gathering focused on a small part of the situation' (23 per cent), and the last mostly consisted of 'Non-detection in a constrained visibility situation' (13 per cent). However, problems of perception can also lead to failures in information processing stages other than the detection stage. As a result of perceptual difficulties we also identified 8 per cent as error of diagnosis and 16 per cent as errors in decision-making (Table 9.1).

As a corollary, we can observe the specificity of failures among PTW riders themselves, depending on whether they are faced with a driver who met a perceptual problem or not. In the case of an interaction, the specific difficulties met by one driver directly had an impact on the error made by the other driver involved. This shows the interactive nature of the failures produced by the participants in any given accident, to such a degree that these accidents can be analysed from the perspective of the 'interaction in the error' (Van Elslande, 2009).

Thus, in accident cases where a PTW perception problem was identified, on the part of the car driver, the PTW driver's failure profile shows an over-representation of prediction failures (42.4 per cent) and diagnosis failures (18.5 per cent). On the other hand, in accident cases without any problems of PTW perception by the car driver, the PTW riders had a detection failure in half of all cases (Figure 9.1). This failure usually corresponds to 'Information gathering focused on a partial component of the situation'. It is therefore the PTW driver who does not detect the other user in these cases. These failures illustrate the fact that the car driver's perceptual problem causes a failure which often leads to behaviour that disturbs the expectations or understanding of the situation by the PTW driver. The PTW perceptual problem therefore has an influence that is both direct, for the type of

Table 9.1   Types of failures involving a perceptual difficulty by car drivers confronted with a PTW

Category of failure	Type of failure	Percentage of failures
Detection	Non-detection in a constrained visibility situation	13.1%
	Information gathering focused on a small part of the situation	23.2%
	Incomplete or hasty information gathering	32.3%
Diagnosis	Erroneous evaluation of a time gap	8.1%
	Violation-error (by automatism)	10.1%
Decision	Violation directed by the context	*3%*
	Deliberate violation of a safety rule	*3%*
	Other failures	*7.2%*
	Total	*100%*
	n	99

error committed by the driver confronted with the PTW, and indirect for the errors made by the PTW drivers in turn.

Furthermore, we assessed the contribution of each factor in the occurrence of a perception failure. These factors could have had a decisive impact on the perception problem or only contributed if they simply compounded the already dysfunctional situation and the perceptual failure may have occurred anyway even if the PTW were absent. Considering this impact, three types of perception problems can be identified.

Firstly, there are perception problems that can be explained by a combination of factors that are specific and non-specific to PTWs, which is what we observe most often in accidents (65.3 per cent of cases). This finding confirms that factors combine and interact to bring about a situation where the other user is not detected. For example, a driver in a complex intersection with heavy traffic and with pressure from behind will be much more vulnerable to the slightest obstruction of visibility, and the rapid information gathering, that would have enabled them to identify a car, may not enable them to perceive a PTW. A powered two-wheeler is never invisible, but it is more vulnerable to perception problems, particularly in certain constrained situations.

The second category involves perceptual problems that are only explained by factors specific to PTW, and if a non-specific factor is identified, it is only a contributing factor rather than the main cause. This group comprises 10.2 per cent

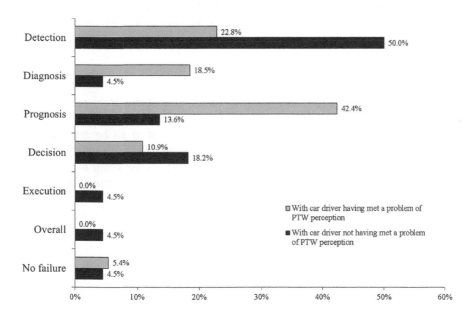

Figure 9.1   Categories of failures for PTW riders confronted with a car driver having met or not met a problem of perception

of our sample of drivers with a PTW perception problem. In these cases, it was presumed that the accident would not have occurred if the PTW had been another type of vehicle. This is the category in which the problem of PTW conspicuity has the most direct impact on the accident.

Finally, there are perception problems which only involve factors that are not specific to PTWs, and if a factor specific to PTWs was identified, it was only a contributing factor. This category includes almost a quarter (24.5 per cent) of the cases analysed. This type of perceptual problem shows that, in nearly one quarter of the cases in which the PTW was not detected, the characteristics (physical, behavioural and cognitive) of the two-wheeler were not instrumental in this misperception. In other words, had the PTW been a four-wheeled vehicle, the accident would have occurred in the same way.

The separation into these three categories provides an indication of the challenges associated with PTW conspicuity, in other words the degree to which being a PTW (compared with other types of vehicles) has an influence on the development of perceptual problems. Thus it appears that in approximately one-quarter of the accidents analysed, where the PTWs were not detected, being faced with a PTW did not have a substantial impact on the perceptual failure. On the other hand, considering all accidents involving a PTW (and not only those with a problem of detection), we can consider that being a PTW had a specific influence on the perceptual problems encountered by the car driver in 57 per cent of cases for all types of functional failures combined. Such results can be compared with the previous work of Williams and Hoffmann (1979) who showed that the low perceptibility of motorcyclists was involved in 64.5 per cent of all cases.

Conspicuity: A Multifaceted Problem

While the literature today is in agreement on the size of the problem involving PTWs, so far no quick-fix solution has emerged from the many studies on the subject. As has been shown by our results, this can first of all be explained by the fact that it is rarely the case that conspicuity is the only explanatory element behind accidents with PTW. PTWs are not invisible on the road, but they are more vulnerable to disturbances therein (for example, lack of attention, obstructed visibility). Thus, from an operational point of view, all measures aimed at improving the safety of interactions between users in general will be beneficial to PTW safety.

Secondly, the problem of conspicuity has both sensory and cognitive origins. Indeed, this problem can be explained by the visual characteristics of PTWs, the sensory capacities of the human perceptual system, the atypical behaviour of PTWs and by the expectations that road users develop. The smaller dimensions of motorcycles on the carriageway is the most commonly mentioned explanatory element (compare with Fulton et al., 1980; Huebner, 1980; Hurt et al., 1981; Thomson, 1980; Williams and Hoffman, 1979; Wulf et al., 1989). Concerning the sensory aspect, Crundall et al. (2008) put forward another hypothesis to explain the differences that they observed in the detection of cars and motorbikes. These authors refer to the hypothesis of spatial frequencies and the theory of global precedence (cf. Hughes et al., 1996). This suggests that information with low spatial frequencies[1] in a visual scene tends to be processed first, and is consequently perceived more quickly than high-frequency information, which corresponds to details which more easily blend into the background. Thus cars, which are wider and have more homogeneous contours and colours, are perceived more frequently by drivers in rapid information gathering. On the other hand, motorbikes, which are narrower and have more complex contours, provide information with high spatial frequency and are not perceived at first glance.

Other authors point out the importance of the cognitive aspect of the detectability problems concerning powered two-wheelers (compare with Hole et al., 1996; Van Elslande and Fouquet, 2007). Thus, according to Hole et al. (1996) the low expectation-level that automobile drivers have concerning motorcyclists is the main reason why they do not perceive them. According to Clarke et al. (2007) this low expectation level is in part related to the rareness of these users in traffic (less than 2 per cent of traffic in France). Van Elslande and Fouquet (2007) also pointed out that the behaviour of PTWs can surprise other road users. Indeed, by deviating from behavioural standards with their manoeuvres, their positioning and the speeds that characterize them, PTWs confound the perceptual strategies that are usually effective for observing other automobile drivers.

The problem of detectability is therefore a rather complex problem that cannot be reduced to the simple fact that PTWs are physically less visible than other vehicles. There are many causes behind the poor detectability of PTWs and these often combine together with the general parameters of the driving context.

References

Clarke, D.D., Ward, P.J., Bartle, C., and Truman, W. (2007). The role of motorcyclist and other driver behaviour in two types of serious accident in the UK. *Accident Analysis and Prevention, 39*, 974–81.

Crundall, D., Humphrey, K., and Clarke, D. (2008). Perception and appraisal of approaching motorcycles at junctions. *Transportation Research Part F, 11*, 159–67.

Donne, G.L. (1990). Research into motorcycle conspicuity and its implementation. Proceeding of the International Congress and Exposition: Society of Automotive Engineers, SAE Technical Paper No. 900749.

---

1   The term spatial frequency refers to the more or less rapid variation of the luminance of a visual pattern in space.

Fulton, E.J., Kirkby, C., and Stroud, P.G. (1980). *Daytime motorcycle conspicuity.* Supplementary Report 625 Leicestershire, UK: Transport and Road Research Laboratory.

Hole, G.J., Tyrrell, L., and Langham, M. (1996). Some factors affecting motorcyclists conspicuity. *Ergonomics, 39*, 946–65.

Huebner, M.L. (1980). ROSTA'S motorcycle visibility campaign – its effect on use of headlights and high visibility clothing, on motorcycle involvement, and on public awareness. Proceedings of the *International Motorcycle Safety Conference*, 715–39.

Hughes, H.C., Nozawa, G., and Ketterle, F. (1996). Global precedence, spatial frequency channels, and the statistics of natural images. *Journal of Cognitive Neuroscience, 8*, 197–230.

Hurt, H.H., Ouellet, J.V., and Thom, D.R. (1981). *Motorcycle accident cause factors and identification of countermeasures.* Report DOT-HS-5-01160, vol. 1 and vol. 2, Washington, D.C.: NHTSA.

MAIDS (2004). *In-depth investigation of motorcycle accidents.* Brussels, ACEM.

ONISR (2010). *Les grandes données de l'accidentologie 2009, caractéristiques et causes des accidents de la route.* Available from http://www.securite-routiere. gouv.fr/

Preusser, D.F., Williams, A.F., and Ulmer, R.G. (1995). Analysis of fatal motorcycle crashes: Crash Typing. *Accident Analysis and Prevention, 27*, 845–52.

Thomson, G.A. (1980). The role frontal motorcycle conspicuity has in road accidents. *Accident Analysis and Prevention, 12*, 165–78.

Van Elslande, P. (2003). Scénarios d'accidents impliquant des deux-roues à moteur : une question d'interaction, In J.M.C. Bastien (Ed.), *Actes des Deuxièmes Journées d'Etude en Psychologie Ergonomique* – EPIQUE'2003, Boulogne Billancourt, 71–83.

Van Elslande, P., and Fouquet, K. (2007). *Analyzing 'human functional failures' in road accidents.* Final report. Deliverable D5.1, WP5 "Human factors". (European TRACE project).

Van Elslande, P. (2009). Erreurs d'interaction, interactions dans l'erreur : spécificité des accidents impliquant un deux-roues motorisé. In P. Van Elslande (Ed.), *Les deux-roues motorisés: Nouvelles connaissances et besoins de recherche.* (Bron: Les Collections de l'Inrets).

Williams, M.J., and Hoffmann, E.R. (1979). Motorcycle conspicuity and traffic accidents. *Accident Analysis and Prevention, 11*, 209–24.

Wulf, G., Hancock, P.A., and Rahimi, M. (1989). Motorcycle conspicuity: An evaluation and synthesis of influential factors. *Journal of Safety Research, 20*, 153–76.

Yuan, W. (2000). The effectiveness of the 'ride-bright' legislation for motorcycles in Singapore. *Accident Analysis and Prevention, 32*, 559–63.

Chapter 10

Applicability of Learner Driver Research to Learner Motorcyclists

Narelle Haworth and Peter Rowden

Centre for Accident Research and Road Safety, Queensland, Australia

Introduction

It is estimated that more than 180,000 motorcycle users are killed in road crashes each year around the world (Naci et al., 2009). Motorcyclists are certainly vulnerable road users, with fatality rates as a function of distance travelled generally about 30 times greater than for car occupants (Johnston et al., 2008; National Center for Statistics and Analysis Research and Development, 2008). The number of motorcycles is increasing in many developed and developing countries (Jamson and Chorlton, 2009; Paulozzi et al., 2007) with consequent increases in fatalities and injuries. The increase in motorcycling means that there are many new riders, who lack experience. Inexperience has been shown to be a major factor in motorcycle crashes (Rutter and Quine, 1996; Mullin et al., 2000) and the common response by governments is to apply graduated licensing principles or systems that have been developed from learner driver research. However, the characteristics of the operators, the vehicles, the importance of road environment factors and usage patterns differ between motorcycles and cars. This chapter will describe these differences and examine the applicability of the findings of learner driver research to learner motorcyclists. It will then discuss the implications for interventions to improve the safety of learner and other motorcyclists.

Comparing the Safety of Learner Motorcyclists and Car Drivers

While fatality and injury rates for learner drivers are typically lower than for those with intermediate licences, this pattern is not found for learner riders. For example, in the Australian State of New South Wales in 2009, learner car drivers comprised 1.1 per cent of drivers in casualty crashes (fatal or injury) while 17.1 per cent of drivers in casualty crashes held a provisional licence (RTA, 2010a). In contrast, learner motorcycle riders made up 12.9 per cent of riders in casualty crashes and riders with provisional licences were involved in 7.6 per cent of casualty crashes.

Some jurisdictions require supervisors of learner car drivers to certify that they have gained a certain number of hours practice whilst under supervision. This requirement is based on the premise that novice car drivers have reduced crash risk with increased hours of supervised practice (Gregersen et al., 2003). Given that learner motorcyclists have a very high crash rate that is certainly not lower than in the first few months of the provisional licence, introducing a requirement for logging hours to increase the number of hours of motorcycle riding experience as an unsupervised learner rider will not have the same benefits as for car drivers. However, a requirement for logged hours of learner riding might discourage some potential learner riders.

Learner Rider Characteristics

The vast majority of learner car drivers are in their teens but learner motorcyclists are typically older and most have extensive driving experience. In many developed countries, the average age of learner riders is mid-thirties (de Rome et al., 2010). Motorcycling also traditionally attracts a far higher proportion of males than females in comparison to car driving, with about 90 per cent of licence holders being male (RTA, 2010b).

While most driving is for transport, the purpose of riding differs markedly among countries and individuals. In the large cities of some European countries most riding is for commuting, while in other developed countries (such as the US and Canada), many motorcycles are leisure vehicles with large engine capacities. In emerging and developing countries motorcycles are largely used as a means of mobility and most are low- and medium-engine capacity motorcycles and scooters (Rogers, 2008).

People are attracted to motorcycling for a variety of reasons including image, the thrill of riding, the feeling of freedom, and to impress others (Watson et al., 2003). The purpose of riding can also affect crash risk, with some research demonstrating that riding for recreation is associated with a higher crash risk than riding for commuting (Haworth et al., 1997). Those high in the personality trait of sensation seeking are often attracted to motorcycling (Horswill and Helman, 2003) and a high propensity for sensation seeking has been shown to predict intentions to engage in risky riding behaviours (Tunnicliff, 2005).

Broughton and Stradling (2005) found that risk taking is an inherent part of enjoyment during riding for some participants (accordingly labelled 'risk seekers' and 'risk acceptors') whilst for others (labelled 'risk averse') the enjoyment of riding came from a sense of freedom rather than risk. This difference highlights the importance of considering fundamental rider motives when attempting to change their behaviour and the understated role of emotions in riding in terms of hedonic motives. These riding motives and motorcyclists' subjective views of risk often do not readily reconcile with expert perceptions of risk (Bellaby and Lawrenson, 2001). The implications of the above findings for learner motorcyclists is that while some will plan to ride responsibly, others are attracted to riding for all the wrong reasons from a road safety point of view. The predominant purpose of riding for each individual is a related factor that influences the safety of all riders and may particularly influence learner motorcyclists' decision to commence riding and their subsequent behaviour.

Motorcycle versus Car Characteristics

Motorcycles and cars differ in terms of occupant protection, ease of supervision, vehicle control and vehicle performance. One of the major contributors to rider injury is their relative lack of protection against impacts with other vehicles, the ground and roadside objects (RoSPA, 2001; Haworth et al., 1997). For this reason, rider injuries are often more severe than those of other road users. For example, approximately 80 per cent of reported motorcycle crashes result in injury or death; a comparable figure for automobiles is about 20 per cent (Motorcycle Safety Foundation, 1999). Errors that might result in minor vehicle damage if committed by a learner driver can result in injury if committed by a learner rider.

The lack of conspicuity, design of roadside barriers, and the use of protective clothing are all additional safety issues related to the vulnerability of motorcyclists. These are known issues that impact on the safety of *all* riders irrespective of licence status; however, potentially some of these impact more on learner riders due to a lack of knowledge of such issues (Rowden, unpublished). The lack of conspicuity of motorcycles can result in other drivers failing to give way to learner riders and placing them in dangerous situations which they may not be able to cope with.

The characteristics of the vehicle mean that the supervision of learner riders is more difficult than for learner drivers. Learner riders are therefore far less likely to receive immediate feedback and instruction from a supervisor rider in order to avoid risky situations or address skill deficiencies. Supervision for motorcyclists is problematic given the increased crash risk associated with carrying a pillion passenger. Since balance and coordination is more difficult with a passenger on a motorcycle, some of the benefits of supervision for novice riders could be achieved by having the supervisor follow the learner on another motorcycle, or closely behind in a car (Mayhew and Simpson, 2001). While it is not expected that the benefits of supervised riding will reduce crash risk per distance travelled as much as it does for learner car drivers, a requirement for supervision could reduce the amount of riding by learner riders because of difficulties in obtaining supervision. Thus, it would be expected to have some road safety benefits.

The challenge of braking effectively to avoid a crash is increased by most motorcycles having separate front and rear braking systems which many learner riders struggle to master. Furthermore, the performance and handling of various types of motorcycles vary greatly. Several studies have reported higher crash rates for sports-style motorcycles and that riders of these motorcycles are younger on average (Harrison and Christie, 2005; Teoh and Campbell, 2010). Scooter riders may have specific training needs due to the handling capabilities of the machine compared to other motorcycles and because most scooters have automatic transmissions.

Vehicle restrictions have been applied to both learner riders and drivers in a range of jurisdictions. Early motorcycle licensing schemes included a maximum engine capacity restriction for learner and/or intermediate licence holders, but there is little evidence of a reduction in crash risk (Langley et al., 2000; Mayhew and Simpson, 1989). Power-to-weight limits have been introduced for learner riders and drivers, with some mixed support for safety benefits for at least learner riders (Elliott et al., 2003; TOI, 2003). Many Australian states have combined power-to-weight and engine capacity restrictions to develop a Learner Approved Motorcycle Scheme (LAMS), which lists motorcycles that may be ridden by learner riders. To date there have been no rigorous evaluations of the effects of these schemes on crash numbers or crash risks.

The Importance of Road Environmental Factors for Motorcycling

Motorcycles have less contact area with the road and are less stable than cars. Hence, recognizing and reacting to road surface irregularities and environmental hazards are more important for riders than for other road users, providing another set of skills to be mastered by learner riders. Common road factors contributing to motorcycle crashes include lack of visibility or obstructions, unclean road or loose material, poor road condition or road markings, and the horizontal curvature of the road (Haworth et al., 1997). Environmental influences such as cold or hot weather, noise, and vibration are all more likely to impact on stress and fatigue states for motorcyclists compared to car drivers (Haworth and Rowden, 2006; Horberry et al., 2008). For learner riders that have not previously been exposed to such factors, there is potential for increased difficulties.

Novice riders have a heightened potential for incorrect responding to hazardous situations due to their inexperience (Wallace et al., 2005). If avoidance manoeuvres such as counter steering and emergency braking have not been perfected there is a greater chance of the rider crashing when a hazardous situation is encountered. Hence, learner riders are potentially more vulnerable than experienced riders due to a lack of skill.

Usage Patterns that Affect Learning and Crash Risk

Many learner riders already own a car and intend to ride the motorcycle for recreation. This can result in some riders gaining little riding experience during the learner period or waiting out periods during which restrictions apply to the type of motorcycle they are allowed to ride, without gaining experience. A survey showed that 18 per cent of riders rode 25 hours or less during the learner phase in New South Wales and that the learner riders had little experience in riding in the rain, at night, in heavy traffic, on winding rural roads or on high speed roads (de Rome et al., 2010). Where there is no minimum duration of the learner phase, the amount of riding as a learner is likely to be much less (Haworth et al., 2010). Where climatic conditions prevent or discourage winter riding, the amount of riding as a learner (and after licensing) may also be reduced.

Implications for Improving Learner Rider Safety

The differences in rider, vehicle and usage patterns between learner riders and drivers have implications for the development of graduated licensing systems for motorcycling and other safety interventions.

The finding that learner riders are often much older and more experienced (as drivers) than learner drivers suggests that graduated licensing systems for motorcycling need to apply irrespective of age, rather than the common practice of age exemptions in graduated driver licensing systems. The implication for rider education and training is that potentially different approaches need to be taken for learner riders who are young and have little or no car driving experience compared to the larger group of learner riders who have extensive on-road experience as a driver. These differences may relate to both content (for example, teaching about road rules) and also to pedagogical styles (adult learning approaches).

The factors related to inexperience that are common to learner car drivers and motorcyclists include the potential for training to contribute to overconfidence, optimism bias and the problem of risk taking. There is a lack of empirical support for the effectiveness of rider training in reducing crash risks above that of informal training or accumulation of experience. Evidence suggests that voluntary motorcycle training programmes do not reduce crash risk (TOI, 2003) and may increase crash risk due, in part, to the increased confidence felt by many riders who have completed training, despite minimal improvements in rider skill. Such riders may therefore take more risks in situations where they lack the skills to safely avoid a crash. Compulsory training through licensing programmes produces a weak but consistent reduction in crashes (TOI 2003) which may be associated with discouraging motorcycling, rather than reducing crash risk. Similarly, a recent paper by Karaca-Mandic and Ridgeway (2010) suggests that graduated driver licensing reduces teen crashes by reducing driving at high-risk times, rather than by reducing crash risk.

The lack of robust evaluations of measures to improve the safety of learner motorcyclists constrains the extent to which conclusions about potential future approaches can be drawn. The available evidence suggests that most of the successful measures have shown reductions in crashes by decreasing riding, rather than by reducing crash risk. There is considerable scope for the improvement of traditional rider training in terms of content, delivery protocols, and the structuring of training within an overall graduated licensing system. Delivering training in stages within a graduated licensing system is important as learners may be more able to integrate information learnt from training once they have had some riding experience as opposed to the learner stage where there is potential for information overload. More focus on higher order factors such as safety attitudes, motivations, and hazard perception offers substantial potential benefits for novice riders and drivers.

References

Bellaby, P., and Lawrenson, D. (2001). Approaches to the risk of riding motorcycles: Reflections on the problem of reconciling statistical risk assessment and motorcyclists' own reasons for riding. *Sociological Review, 49*(3), 368–89.

Broughton, P.S., and Stradling, S.G. (2005). Why ride powered two-wheelers? In *Proceedings from Behavioural Research in Road Safety: Fifteenth seminar.* London: Department of Transport.

de Rome, L., Ivers, R., Haworth, N.L., Heritier, S., Fitzharris, M., and Du, W. (2010). A survey of novice riders and their riding experience prior to licensing. *Transportation Research Record*, Journal of the Transportation Research Board of the National Academies, No. 2194, 75–81.

Elliott, M., Baughan, C.J., Broughton, J., Chinn, B., Grayson, G.B., Knowles, J., Smith, L.R., and Simpson, H. (2003). Motorcycle safety: A scoping study. *TRL Report 581.* Crowthorne: Transport Research Laboratory.

Gregersen, N.P., Nyberg, A., and Berg, H-Y. (2003). Accident involvement among learner drivers: An analysis of the consequences of supervised practice. *Accident Analysis and Prevention, 35*, 725–30.

Harrison, W.A., and Christie, R. (2005). Exposure survey of motorcyclists in New South Wales. *Accident Analysis and Prevention, 37*, 441–51.

Haworth, N., and Rowden, P. (2006). Fatigue and motorcycling: Is there an issue? Proceedings of the *Australasian Road Safety Research, Policing and Education Conference*, Gold Coast, Queensland, 25–6 October 2006. Brisbane: Queensland Transport (CD-ROM).

Haworth, N., Rowden, P.J., and Schramm, A.J. (2010). A preliminary examination of the effects of changes in motorcycle licensing in Queensland. Paper presented at the *Australasian Road Safety Research, Policing and Education Conference*, Canberra, 31 August – 2 September 2010.

Haworth, N., Smith, R., Brumen, I., and Pronk, N. (1997). Case-control study of motorcycle crashes (*Contract report CR174*). Canberra: Federal Office of Road Safety.

Horberry, T., Hutchins, R., and Tong, R. (2008). Motor cycle rider fatigue: A review. *Research Report Number 8, Great Britain.* London: Department for Transport.

Horswill, M.S., and Helman, S. (2003). A behavioural comparison between motorcyclists and a matched group of non-motorcyclists car drivers: Factors influencing accident risk. *Accident Analysis and Prevention, 35*, 589–97.

Jamson, S., and Chorlton, K. (2009). The changing nature of motorcycling: Patterns of use and rider characteristics. *Transportation Research Part F, 12*, 335–46.

Johnston, P., Brooks, C., and Savage, H. (2008). Fatal and serious road crashes involving motorcyclists. *Road safety research and analysis report: Monograph 20*. Canberra: Australian Department of Infrastructure, Transport, Regional Development and Local Government.

Karaca-Mandic, P., and Ridgeway, G. (2010). Behavioral impact of graduated driver licensing on teenage driving risk and exposure. *Journal of Health Economics, 29*, 48–61.

Langley, J.D., Mullin, B., Jackson, R., and Norton, R. (2000). Motorcycle engine size and risk of moderate to fatal injury from a motorcycle crash. *Accident Analysis and Prevention, 32*(5), 659–63.

Mayhew, D.R., and Simpson, H.M. (1989). *Motorcycle engine size and traffic safety*. Ottawa, Canada: Traffic Injury Research Foundation.

Mayhew, D.R., and Simpson, H.M. (2001). *Graduated licensing for motorcyclists*. Ottawa, Canada: Traffic Injury Research Foundation.

Motorcycle Safety Foundation. (1999). *Motorcycle Safety*. Irvine, CA: MSF. Available from www.msf-usa.org/

Mullin, B., Jackson, R., Langley, J., and Norton, R. (2000). Increasing age and experience: Are both protective against motorcycle injury? A case control study. *Injury Prevention, 6*, 32–5.

Naci, H., Chisholm, D., and Baker, T.D. (2009). Distribution of road traffic deaths by road user group: A global comparison. *Injury Prevention, 15*, 55–9.

National Center for Statistics and Analysis (2008). *Traffic Safety Facts 2006 Data – Motorcycles*, updated March 2008. Available from http://wwwnrd.nhtsa.dot.gov/ Pubs/810620.PDF (DOT HS 810 806).

Paulozzi, L.J., Ryan, G.W., Espitia-Hardeman, V.E., and Xi, Y. (2007). Economic development's effect of road-transport related mortality among different types of road users: A cross-sectional international study. *Accident Analysis and Prevention, 39*, 606–17.

Rogers, N. (2008). Trends in Motorcycles Fleet Worldwide. Presentation to Joint OECD/ITF Transport Research Committee Workshop on Motorcycling Safety. Available from http://www.internationaltransportforum.org/jtrc/safety/ Lillehammer2008/Lillehammer08Rogers.pdf/

Royal Society for the Prevention of Accidents (RoSPA) (2001). *Motorcycling Safety Position Paper*. Available from http://www.rospa.co.uk/pdfs/road/ motorcycle.pdf

RTA (2010a). *Road Traffic Crashes in New South Wales. Statistical Statement for the year ended 31 December 2009*. NSW Centre for Road Safety, Roads and Traffic Authority of New South Wales.

RTA (2010b). *New South Wales Driver and Vehicle Statistics 2010 (Preliminary)*. Available from http://www.rta.nsw.gov.au/publicationsstatisticsforms/ downloads/registration/year_book_2010_preliminary.xls/

Rutter, D.R., and Quine, L. (1996). Age and experience in motorcycle safety. *Accident Analysis and Prevention, 28*, 15–21.

Teoh, E.R., and Campbell, M. (2010). *Role of motorcycle type in fatal motorcycle crashes*. Arlington, VA: Insurance Institute for Highway Safety.

TOI. (2003). *Motorcycle safety – a literature review and meta-analysis of countermeasures to prevent accidents and reduce injury*. (English summary).

Tunnicliff, D. (2005). *Psychosocial factors contributing to motorcyclists' intended riding style: An application of an extended version of the theory of planned behaviour*. Unpublished Master's thesis. Brisbane: Queensland University of Technology.

Wallace, P., Haworth, N., and Regan, M. (2005). Best training methods for teaching hazard perception and responding by motorcyclists. *Report 236*. Melbourne: Monash University Accident Research Centre.

Watson, B. Tay, R., Schonfeld, C., Wishart, D., Lang, C., and Edmonston, C. (2003). *Short-term process and outcome evaluation of Q-RIDE*. Unpublished report to Queensland Transport. Brisbane: Centre for Accident Research and Road Safety, Queensland.

Chapter 11

Influence of Cognitive Bias on Young Cyclists' Road Crossing Intentions at Non-Signalized Intersections

Yasunori Kinosada and Shinnosuke Usui
Graduate School of Human Sciences, Osaka University, Japan

Introduction

Road traffic statistics from the National Police Agency (2010) indicate that the number of traffic accident fatalities in Japan has been decreasing; however, the reduction rate differs by accident type. The number of car-related accident deaths has decreased by 58.68 per cent in the past decade in Japan, whereas bicycle-related deaths have only decreased by 32.66 per cent. Bicycle riding is an environmentally friendly mode of transportation and it is beneficial to the riders' health. In Japan, bicycle riding is very popular, especially for shopping and commuting. Therefore, it is important to determine the cause of bicycle-related accidents and to implement effective interventions.

Cyclists are quite vulnerable on the road; thus, they are expected to behave safely in heavy traffic (for example, with cars and motorcycles). However, accident data suggest that cyclists violated traffic laws in about 70 per cent of bicycle-related accidents. Cyclists' unsafe behaviour may be related to their distorted judgements (National Police Agency, 2010). Räsänen and Summala (1998) found that when cyclists have the right of way at non-signalized intersections, they often expect oncoming cars to yield. According to Räsänen and Summala, such misplaced expectations resulted in bicycle-car collisions. Taylor and Brown (1988) stated 'Prior expectations and self-serving interpretations weigh heavily into the social judgement process'. Thus, it is possible that expectations and self-serving interpretations are related to bicycle accidents.

Using questionnaires based on the theory of planned behaviour (Ajzen, 1991), this study sought to verify the influence of expectations and self-serving interpretations on cyclists' unsafe intentions to cross non-signalized intersections. The theory of planned behaviour presumes that intention triggers behaviour. Intentions are determined by attitude, subjective norms, and perceived behavioural control. Attitude refers to an individual's positive or negative evaluation of the particular behaviour under investigation. Subjective norms are the perceived social pressure from important others to commit, or not commit a particular behaviour. Perceived behavioural control reflects how easy an individual believes it is to commit or refrain from the behaviour. The theory of planned behaviour has been applied to predicting such unsafe traffic behaviours as: speeding, overtaking, and drink driving (Parker et al., 1992). Furthermore, Evans and Norman (1998; 2003) used this theory to predict pedestrian accidents. Although this theory has not been used to test cyclists' unsafe behaviour in general, it has been used to predict bicycle helmet use (Quine et al., 2001). Traffic statistics from the National Police Agency (2010) indicate that more than half of bicycle accident fatalities involve head injuries in Japan. Wearing a helmet may lessen the severity of injuries in an accident, but it cannot prevent the accident. Predicting cyclists' unsafe crossing intentions using the theory of planned behaviour may clarify the causes of bicycle accidents.

According to Ajzen (1991) 'the theory of planned behaviour is, in principle, open to the inclusion of additional predictors if it can be shown that they capture a significant proportion of the variance in intention or behaviour after the theory's current variables have been taken into account'. In this study, variables measuring expectations and self-serving interpretations were added. The first additional variable was the expectation that cars would yield (Räsänen and Summala, 1998). In addition, variables involving self-serving interpretations, self-serving bias, and unrealistic optimism were included. Self-serving bias is a type of attribution error. According to Miller and Ross (1975) 'people accept more responsibility for their successes than for their failures'. In addition, Weinstein (1980) stated 'People tend to think they are invulnerable. They expect others to be victims of misfortune, not themselves. Such ideas imply not merely a hopeful outlook on life, but an error in judgement that can be labelled *unrealistic optimism*'.

Method

Respondents

A total of 148 college students (41.89 per cent males) answered the questionnaire. The group's average age was 20.47 years ($SD = 1.07$; range = 19 to 24), and 52.70 per cent of them held automobile or motorcycle driving licences, or both. Of these respondents, 66.89 per cent used their bicycles at least once a week.

Procedure and Design of the Questionnaire

Two-thirds of the questionnaires were distributed at the end of college classes. The rest were handed to respondents individually. Respondents were instructed to imagine the situation depicted by a scenario and a picture illustrating a car and a bicycle approaching a non-signalized intersection (see Figure 11.1). The car was approaching from a crossroad to the right of the cyclist. Eight situations were depicted with three factors: right of way (cyclist moving on a major or minor road), distance (far from or near the approaching car on the crossroad), and psychological state (whether the cyclist was in a hurry or was calm).

The right of way of the road was presented both by scenario and picture, to help inform respondents who did not hold driving licences. When the cyclist was on a major road, there was a stop sign and a halt line at the crossroad.

The distance from the approaching car was manipulated in the pictures. Two patterns of the position of the approaching car (far or near) were set on both major and minor roads. On a major road, the physical distances between the cyclist and the car were 6.5 m and 9 m, while on a minor road they were 16 m and 24 m.

Figure 11.1   Example of picture presented in a minor–near situation

The psychological state was described in the scenario. According to preliminary research, the end-of-term exam is important to college students, meaning it could arouse a sense of urgency. The following (hurrying) scenario was presented to the respondents.

> Today you have an important end-of-term exam. However, you overslept and are rushing to school by bicycle. When you reach the intersection, you see an oncoming car; but you try to cross without stopping.

Preliminary research showed that attending classes was viewed as less important to college students. The following (calm) scenario was presented to the respondents.

> Today you have classes at college. You woke up early, so you decided to go to college by bicycle, instead of taking the bus as you usually do. When you reach the intersection, you see an oncoming car, but you try to cross without stopping.

The psychological state was the only between-subjects factor, so each questionnaire contained four situations.

Measures

For each situation, respondents answered items about their intention to cross the intersection, components of the theory of planned behaviour (attitude, subjective norm, and perceived behavioural control), and additional items about their expectations and self-serving interpretations (expectation that the cars would yield, self-serving bias, and unrealistic optimism) on a seven-point scale, with higher scores indicating stronger intentions.

> *Intention to Cross the Intersection*: Intention was rated by how much they wanted to cross the intersection (1 = Very weakly to 7 = Very strongly).
> *Components of the Theory of Planned Behaviour*: Four items focused on attitude. Two were related to behavioural beliefs regarding the likelihood of the results (involvement in an accident and getting to college earlier; rated from 1 = Very unlikely to 7 = Very likely) for each crossing scenario. The other two questions were about evaluating the potential outcomes described above (rated from 1 = Very bad to 7 = Very good). The attitude score was the averaged products of the behavioural beliefs and the evaluation of each outcome.

There were eight items measuring subjective norms, four of which measured normative beliefs. Respondents were asked to what degree important others would want them to cross (1 = Very weakly to 7 = Very strongly). The other four items measured motivation to comply with the wishes of their important others (1 = Very weakly to 7 = Very strongly). The normative beliefs and motivation to comply were multiplied for the important others (parents, driver, another cyclist behind the participant, and the teacher of the exam or class) and these four products were averaged to form the subjective norms score.

Ajzen (1991) suggested that perceived behavioural control can be measured using two items: the factors that facilitate or distract the behaviour, and the power of such factors. However, many studies (for example, Godin et al., 1992; Sparks et al., 1995) have used a single item: the degree of difficulty to cross the road, as depicted in the scenario (1 = Very difficult to 7 = Very easy). For ease of answering, this study also used a single question.

> *Expectation and Self-serving Interpretations*: Expectation that cars would yield was investigated using a single item. Respondents were asked to indicate to what degree they expected the automobile driver to yield in each situation (1 = Not at all to 7 = A great deal). For the item measuring self-serving bias, respondents were asked to rate their degree of responsibility if an accident resulted from the attempted crossing (1 = A great deal to 7 = Not at all). Unrealistic optimism involved rating the possibility of their being involved in an accident compared with that possibility for the average college student (1 = Very likely to 7 = Very unlikely).

The scores of the three above items were averaged and labelled 'cognitive bias' to investigate the comprehensive effect of expectations and self-serving interpretations on intention.

Analysis

Intention is the most proximal trigger of behaviour according to the theory of planned behaviour; thus, it is necessary to confirm that each of these three factors (right of way, distance, and psychological state) influenced intention to perform the behaviour. A three-way analysis of variance (right of way × distance × psychological state) was conducted, with intention as the dependent variable.

Next, hierarchical regression analyses, with intention as the dependent variable, were performed for each situation. Using the forced entry method, independent variables were divided into three steps. In step one the three demographic variables (gender, with or without a driving licence, and do or do not use a bicycle) were entered. The demographic variables were converted to categorical data (male = 1, female = 0; with a driving licence = 1, without a driving licence = 0; uses a bicycle = 1, does not use a bicycle = 0). In step two the components of the theory of planned behaviour were added. At step three, the influence of cognitive bias was investigated by entering this variable (which was an average score of the three items measuring expectations and self-serving interpretations).

Results

The results of the three-way analysis of variance indicated that the interaction between right of way and car distance was significant ($F[1, 146] = 4.34, p < .05$). A post hoc test indicated that when the approaching car was farther away, intentions to cross were stronger, regardless of whether they had the right of way ($ps < .001$). Also, regardless of distance, intentions to cross were stronger when cycling on a major road, rather than on a minor road ($ps < .001$). However, surprisingly psychological state did not affect intention.

Following the analysis of variance, the psychological states were combined to further analyse the right of way and car distance. Means and standard deviations of all variables are presented in Table 11.1.

For each situation, hierarchical regression analyses were performed. The three variables measuring expectation and self-serving interpretations (expectation that cars would yield, self-serving bias, and unrealistic optimism) had acceptable internal consistency in every situation. Cronbach's alpha coefficients were 0.71 in the major-far situation, 0.72 in the major-near situation, 0.68 in the minor-far situation, and 0.70 in the minor-near situation. Therefore, these items were averaged and combined for each situation and labelled 'cognitive bias'. The cognitive bias variable was entered as an independent variable at the third step of the hierarchical regression analyses.

The correlations between intention and the independent variables are presented in Table 11.2 and the results of the hierarchical regression analyses are presented for each situation in Tables 11.3, 11.4, 11.5 and 11.6.

Table 11.1   Descriptive statistics of the variables in each situation

	Right of Way			
	Major		Minor	
	Distance		Distance	
Variables (Range of scores)	Far	Near	Far	Near
Intention (1-7)	4.72 (1.94)	4.03 (2.13)	3.50 (2.18)	2.40 (1.75)
Attitude (1-49)	14.32 (5.27)	13.57 (5.26)	13.80 (5.59)	13.34 (5.48)
Subjective norm (1-49)	13.87 (7.93)	12.50 (6.84)	11.36 (6.56)	8.90 (4.37)
Perceived behavioural control (1-7)	4.85 (1.62)	4.18 (1.79)	3.74 (1.90)	2.93 (1.59)
Cognitive bias (1-7)	4.82 (1.17)	4.42 (1.38)	3.66 (1.41)	3.09 (1.19)

Note: Standard deviations are in parentheses

Table 11.2   Correlation coefficients between intentions and the independent variables

	Situations			
	Major-Far	Major-Near	Minor-Far	Minor-Near
Gender	.22 **	.14 *	.17 *	.10
Driving licence	.17 *	.22 **	-.12	-.20 **
Usage of bicycle	-.08	.04	-.17 *	-.28 ***
Attitude	.17 *	.22 **	.34 ***	.30 ***
Subjective norm	.39 ***	.45 ***	.41 ***	.33 ***
Perceived behavioural control	.72 ***	.78 ***	.76 ***	.70 ***
Cognitive bias	.59 ***	.72 ***	.64 ***	.50 ***

***: $p < .001$, **: $p < .01$, *: $p < .05$

Major-Far Situation

Whether or not the individual used a bicycle did not correlate significantly with intention to cross the intersection (Table 11.2). The results of the hierarchical regression analysis are presented in Table 11.3. At step one only gender was significant, meaning that males were more likely to cross the intersection when they were riding on a major road and the oncoming car was distant. In step two, the subjective norm and perceived behavioural control were significant predictors of intention. However, in step three, the effect of the subjective norm was no longer significant, while the cognitive bias variable was significant. Having a strong cognitive bias increased the participants' unsafe intentions to cross the intersection. For each step, the increased proportion of variance explained (ΔR^2) was significant.

Table 11.3   Hierarchical regression for the major–far situation

Step	Independent variables	β1	β2	β3	F	Adj R^2	ΔR^2
	Gender	.20 **	.06	.06			
1	Driving licence	.12	.04	.02	3.67 *	.05	
	Usage of bicycle	-.09	.01	.01			
	Attitude		.04	.03			
2	Subjective norm		.14 *	.07	28.39 ***	.53	.48 ***
	Perceived behavioural control		.65 ***	.54 ***			
3	Cognitive bias			.25 ***	28.28 ***	.57	.04 ***

***: $p < .001$, **: $p < .01$, *: $p < .05$
β: standardized partial regression coefficient at each step
Adj R^2: adjusted coefficient of determination
ΔR = change in R^2

Major-Near Situation

Whether or not the individual used a bicycle did not correlate significantly with intention to cross the intersection (Table 11.2). The results of the hierarchical regression are presented in Table 11.4. In step one the respondents who have driving licences were more likely to cross the intersection when they were riding on a major road and the oncoming car was close. In step two and three, perceived behavioural control was the only significant predictor of intention to cross. However, at step three the cognitive bias variable was again a significant predictor of intention to cross the intersection. For each step, the increased proportion of variance explained (ΔR^2) was significant.

Table 11.4   Hierarchical regression for the major–near situation

Step	Independent variables	β1	β2	β3	F	Adj R^2	ΔR^2
	Gender	.10	.01	.00			
1	Driving licence	.20 *	.11 *	.11 *	2.93 *	.04	
	Usage of bicycle	.03	.00	.01			
	Attitude		.03	.04			
2	Subjective norm		.05	.00	39.16 ***	.61	.57 ***
	Perceived behavioural control		.73 ***	.53 ***			
3	Cognitive bias			.33 ***	42.57 ***	.66	.06 ***

***: $p < .001$, *: $p < .05$
β: standardized partial regression coefficient at each step
Adj R^2: adjusted coefficient of determination
ΔR = change in R^2

Minor-Far Situation

Whether the participants held driving licences or not did not correlate with intention (Table 11.2). In step one of the hierarchical regression, all demographic variables were significant predictors of intentions to cross the intersection (Table

11.5). Male respondents, those who did not use a bicycle and those who did not have a driving licence were more likely to intend to cross the intersection when they were riding on a minor road and the oncoming car was distant. In steps two and three, perceived behavioural control was a significant predictor of intentions. At step three, cognitive bias was found to significantly predict unsafe crossing intentions. Again, the change in the proportion of variance explained (ΔR^2) was significant for each step.

Table 11.5   Hierarchical regression for the minor–far situation

Step	Independent variables	$\beta 1$	$\beta 2$	$\beta 3$	F	Adj R^2	ΔR^2
	Gender	.22 *	.09	.09			
1	Driving licence	-.16 *	.04	.05	4.46 **	.07	
	Usage of bicycle	-.17 *	-.06	-.05			
	Attitude		.11	.12 *			
2	Subjective norm		.07	.03	36.18 ***	.59	.52 ***
	Perceived behavioural control		.68 ***	.58 ***			
3	Cognitive bias			.18 *	32.67 ***	.60	.01 ***

***: $p < .001$, **: $p < .01$, *: $p < .05$

β: standardized partial regression coefficient at each step

Adj R^2: adjusted coefficient of determination

ΔR = change in R^2

Minor-Near Situation

Gender did not significantly correlate with intention to cross the intersection (Table 11.2). The results of the hierarchical regression are presented in Table 11.6. At step one all demographic variables were significant, but at step two only whether participants use a bicycle or not remained significant. Of the TPB variables, only perceived behavioural control predicted unsafe crossing intentions. From step one to two, the increased proportion of variance explained (ΔR^2) was significant. However, in step three cognitive bias did not significantly predict unsafe crossing intentions, meaning that the increased proportion of variance explained (ΔR^2) was not significant.

Table 11.6   Hierarchical regression for the minor–near situation

Step	Independent variables	$\beta 1$	$\beta 2$	$\beta 3$	F	Adj R^2	ΔR^2
	Gender	.17 *	.08	.07			
1	Driving licence	-.23 *	-.06	-.04	8.01 ***	.13	
	Usage of bicycle	-.29 ***	-.13 *	-.13 *			
	Attitude		.10	.08			
2	Subjective norm		.11	.10	27.09 ***	.52	.39 ***
	Perceived behavioural control		.59 ***	.53 ***			
3	Cognitive bias			.13	24.09 ***	.52	.01

***: $p < .001$, *: $p < .05$

β: standardized partial regression coefficient at each step

Adj R^2: adjusted coefficient of determination

ΔR = change in R^2

Discussion

Constructing Situations

The three-way analysis of variance indicated that psychological state did not have a significant effect on intention to cross the intersection, which may be due to two problems. Firstly, the objects in the scenarios were established on the basis of preliminary research undertaken by the authors. However, the importance of an end-of-term exam and college classes may differ for each respondent. Secondly, the traffic situations that the cyclist faced in the scenarios were not stressful. For example, in the scenarios Evans and Norman (1998) used, unsafe crossings were depicted as situational dilemmas. For example, where crossing facilities were unavailable, or where crossing facilities were located in a different direction to that which the participant wanted to go. In the scenarios used in this study, a single approaching car was not disruptive enough to engender a feeling that enough time would be lost in order to distinguish between the two scenarios.

The interaction between right of way and distance was confirmed. Thus, given the non-significant effect of the psychological state, constructing four situations with the remaining two factors allowed better differentiation between the situations.

Predicting Crossing Intentions in Each Situation

Some of the respondents' characteristics were significant predictors of crossing intentions in each situation. However, the proportion of variance explained was so low ($R^2 = 0.04$ to 0.13) that the models in step one were poor predictors of crossing intentions.

The components of the theory of planned behaviour explained a substantial proportion of the variance in crossing intentions, even after controlling for the effects of the demographic variables ($R^2 = 0.39$ to 0.57). The proportions of the variance explained here are higher than those in other studies using the theory of planned behaviour to explain pedestrians' unsafe crossing intentions (for example, Evans and Norman, 1998; 2003). This therefore highlights the utility of the theory of planned behaviour for predicting cyclists' unsafe crossing intentions.

Among the components of the theory of planned behaviour, perceived behavioural control was a significant predictor of unsafe intentions in all situations. Regardless of whether they had the right of way, or the distance from the approaching car, feeling at ease when crossing potentially hazardous intersections is very dangerous. Attitude and subjective norms also influenced unsafe intentions in some situations. However, as reported by Ajzen (1991), the effect of these two variables was less than that of perceived behavioural control. Ajzen (2006) proposed that 'important others are generally perceived to approve of desirable behaviour and disapprove of undesirable behaviour'. The situations presented in this study were generally much more risky than those used in previous studies. Therefore, with this in mind, the smaller effect of subjective norms would have been expected. Similarly, the weaker effect of attitude was within the scope of the assumption because the evaluation of more risky crossing behaviour should be lower. To investigate the influence of attitude and subjective norm on unsafe crossing, further research using less risky behaviours is needed.

The addition of the cognitive bias variable increased the explained variance in intentions in most situations ($\Delta R^2 = 0.01$ to 0.06). It is notable that the effect of cognitive bias was not significant in the Minor-Near situation, but that cognitive bias was positively correlated with intention ($r = 0.50$). According to the theory of planned behaviour, the components of the theory of planned behaviour are supposed to explain most of the variance in intentions by step two. Thus, although the increases in R^2 were quite low from step two to three, the increase in explained variance was over and above that explained by the demographic variables as well as the components of the theory of planned behaviour. Furthermore, when the cognitive bias variable was added, the effect of subjective norms became insignificant in the Major-Far situation. This result implies that the influence of cognitive bias on unsafe crossing intentions is stronger than subjective norms when a car is approaching non-signalized intersections.

The scores of the three items measuring expectations and self-serving interpretations were averaged, and their individual effects on cyclists' unsafe intentions were not examined. According to Räsänen and Summala (1998), cyclists who have the right of way at non-signalized intersections tend to expect oncoming cars to yield. However, in this study, the influence of cognitive bias was not limited to situations where the cyclist had the right of way. Thus, it is possible that cyclists expect cars to yield regardless of right of way. Further research is needed to clearly understand the relationship between the individual concepts of cognitive bias and the characteristics of an intersection.

Conclusion

This study confirmed not only the usefulness of the theory of planned behaviour for predicting cyclists' unsafe crossing intentions, but also the finding that cyclists' crossing intentions, when a car is approaching non-signalized intersections, were unsafely influenced by cognitive bias. Even when it was clear that a car was approaching, cognitive bias promoted unsafe crossing intentions. It is particularly noteworthy that, in contrast to the findings of Räsänen and Summala (1998), the influence of cognitive bias may not be limited to situations in which cyclists have the right of way. Even when cyclists were riding on a minor road, cognitive bias predicted unsafe crossing intentions in the Minor-Far situation. Although the mean score for crossing intentions in this situation was low, this highlights the importance of considering cyclists' cognitive bias to predict their unsafe intentions. Consequently, educational interventions to lower cognitive bias may help reduce cyclists' unsafe crossing behaviour.

In future research, the influence of cognitive bias on unsafe crossing intentions should be investigated by including other types of distorted judgements. Likewise, it needs to be ascertained whether intentions predicted by the components of the theory of planned behaviour and cognitive bias are linked to actual behaviour.

References

Ajzen, I. (1991). The theory of planned behaviour. *Organizational Behaviour and Human Decision Processes, 50*, 179–211.

Ajzen, I. (2006). *Constructing a tpb questionnaire: Conceptual and methodological considerations.* Retrieved November 10, 2010, from http://www.unibielefeld.de/ikg/zick/ajzen%20construction%20a%20tpb%20questionnaire.pdf

Evans, D., and Norman, P. (1998). Understanding pedestrians' road crossing decisions: An application of the theory of planned behaviour. *Health Education Research, 13*(4), 481–9.

Evans, D., and Norman, P. (2003). Predicting adolescent pedestrians' road-crossing intentions: An application and extension of the theory of planned behaviour. *Health Education Research, 18*(3), 267–77.

Godin, G., Valois, P., Lepage, L., and Desharnais, R. (1992). Predictors of smoking behaviour: An application of Ajzen's theory of planned behaviour. *British Journal of Addiction, 87*, 1335–43.

Parker, D., Manstead, A.S.R., Stradling, S.G., Reason, J.T., and Baxter, J.S. (1992). Intention to commit driving violations: An application of the theory of planned behaviour. *Journal of Applied Psychology, 77*(1), 94–101.

Miller, D. T., and Ross, M. (1975). Self-serving biases in attribution of causality: Fact or fiction? *Psychological Bulletin, 82*, 213–225.

National Police Agency (2010). *Traffic accident occurrence in 2009.* Retrieved August 31, 2010, from http://www.e-stat.go.jp/SG1/estat/List.do?lid=000001062201

Quine, L., Rutter, D. R., and Arnold, L. (2001). Persuading school-age cyclists to use safety helmets: Effectiveness of an intervention based on the Theory of Planned Behaviour. *British Journal of Health Psychology, 6*, 327 – 345.

Räsänen, M., and Summala, H. (1998). Attention and expectation problems in bicycle-car collisions: An in-depth study. *Accident Analysis and Prevention, 30*(5), 657–66.

Sparks, P., Shepherd, R., Wieringa, N., and Zimmermanns, N. (1995). Perceived behavioural control, unrealistic optimism and dietary change: An exploratory study. *Appetite, 24*(3), 243–55.

Taylor, S.E., and Brown, J.D. (1988). Illusion and well-being: A social psychological perspective on mental health. *Psychological Bulletin, 103*, 193–210.

Weinstein, N.D. (1980). Unrealistic optimism about future life events. *Journal of Personality and Social Psychology, 39*(5), 806–20.

PART IV
Hazard Perception and Risk

Chapter 12

Driver Fatigue:
The Perils of Vehicle Automation

Gerald Matthews, Catherine E. Neubauer, Dyani J. Saxby
and Lisa K. Langheim
Department of Psychology, University of Cincinnati, USA

The Nature of Driver Fatigue

Driver stress and fatigue are well known to be potentially dangerous during vehicle driving (Hitchcock and Matthews, 2005). The effects of fatigue are particularly well documented. Evidence for the dangers of fatigue has come from various sources. These include analyses of causal factors in crashes, video observation of driving, company records, and experimental studies using driving simulators (see Williamson et al., 2011, for a review). However, the term 'fatigue' is rather vague and notoriously hard to define. It may be seen as a subjective state characterized by tiredness (Matthews et al., forthcoming a), as a loss of task motivation and unwillingness to commit effort (Brown, 1997), or as a biologically driven need for rest (Williamson et al., 2011).

There are multiple sources of fatigue, including the homeostatically controlled need for sleep that increases with hours of wakefulness, circadian rhythms and task-induced fatigue associated with prolonged monotonous work (Matthews et al., 2000). Broadly, all these sources may lead to impairments in attention and information processing, but the specific mechanisms of driving performance change may differ. In this chapter, we focus especially on task-induced fatigue that may lead to performance deficits in the wakeful, but tired, driver. A recent study has shown that, in short-haul truck drivers mainly working day shifts, workload factors predict the frequency of fatigue (Friswell and Williamson, 2008). Thus, there is a need for research on workload and fatigue to complement the larger applied literature on sleep loss effects.

Fatigue covers a variety of different mental states, which may differ in their implications for performance. Tiredness is a core element of fatigue (Matthews et al., forthcoming a). Other symptoms, which are sometimes, but not always, present include loss of motivation, mind-wandering, emotional distress and various forms of physical discomfort (Matthews and Desmond, 1998). In addition, fatigue is often associated with attempts at coping, such as playing the radio to add stimulation. Drivers typically try to regulate their mental state, motivated both by safety concerns and by needs to maintain personal comfort (Fairclough, 2001).

Thus, fatigue may be treated as a multidimensional state. In this chapter, our analysis is based on a three-dimensional model of subjective states proposed by Matthews et al. (2002). The model differentiates (1) task engagement (energy, motivation, and concentration), (2) distress (tension, unpleasant mood, low confidence) and (3) worry (self-focused attention, low self-esteem, intrusive thoughts). Task-induced fatigue is most typically expressed as low task engagement (Matthews et al., 2010), and quite commonly by elevated distress also (Matthews and Desmond, 2002). Desmond and Matthews (2009) also noted a modest elevation of intrusive thoughts in a study of long-distance truckers, although worry is not a prominent symptom in simulator studies (Matthews and Desmond, 2002). A more elaborate taxonomy for fatigue states, that includes dimensions for physical fatigue, self-appraisals and coping strategies, has been presented by Hitchcock and Matthews (2005) and Matthews et al. (forthcoming a).

Workload Interventions

Theoretical accounts of task-induced fatigue typically emphasize workload management (Hockey, 2011; Matthews and Desmond, 2002). The fatigued performer is prone to adopt lower goals for performance to reduce cognitive load (Hockey, 2011). In the transportation context, the fatigued driver may tend to adopt a largely reactive strategy that minimizes active attempts to maintain safety, while maintaining some level of alertness for traffic-related events (Matthews and Desmond, 2002).

Driver fatigue may also be understood within transactional theories of stress and emotion (Lazarus, 1999) that see 'stress' as a process associated with dynamic self-evaluation and coping with external demands. Matthews (2002; Desmond and Matthews, 2009) suggest that driver fatigue is controlled by reluctance to initiate task-focused coping that leads to a withdrawal of effort from task-related processing. Drivers' performance strategies may become increasingly reactive, as they seek to reduce cognitive load and to invest some minimal level of effort for maintaining safety.

It is important to differentiate two separate roles for workload in the fatigue process. First, workload may play a role in generating fatigue, as a demand factor that elicits efforts at coping. The role of workload is supported by studies showing that prolonged high workloads tend to impair sustained attention (Warm et al., 2008), and field-based studies linking fatigue to high workloads (Friswell and Williamson, 2008). Second, workload may *moderate* the impact of fatigue on performance. We will discuss whether cognitive underload and/or overload may accentuate fatigue-related impairment.

The role of workload suggests a role for vehicle technology in fatigue management. Vehicle technologies that automate functions including speed regulation and hazard detection are becoming increasingly prevalent. Technological advance may eventually culminate in fully automated highway systems. Current systems such as adaptive cruise control (ACC) tend to reduce driver workload, and so should mitigate the fatigue and stress associated with high workloads.

These hopes are not entirely misplaced. In a simulator study, Stanton and Young (2005) confirmed that use of ACC reduced driver stress and workload. However, automation may not provide a remedy for driver fatigue and stress. Young and Stanton (2007) point out that underload is potentially dangerous. They confirmed, in a simulator study, that use of automation led to increased braking response times, suggesting a potentially dangerous loss of situation awareness. Matthews and Desmond (2002) found that fatigue was detrimental to driver performance, indexed by a measure of vehicle control, only in low workload conditions. They attributed this finding to a breakdown in effort-regulation when the task is perceived as undemanding.

This chapter will summarize evidence from recent studies on three key issues: (1) whether automation induces or exacerbates driving fatigue and stress, (2) whether these fatigue responses impair driver performance and safety, and (3) the role of individual differences in the fatigue response. We will focus especially on recent studies from our laboratory that examine fatigue response to simulations of automated driving.

Effects of Automated Driving on Fatigue Response

How does use of automation influence fatigue and stress response? A conceptual basis for addressing the issue is provided by the differentiation of two qualitatively different forms of fatigue (Desmond and Hancock, 2001). Active fatigue is elicited when the driver must produce frequent control responses under prolonged high workload (for example, driving on a busy freeway). By contrast, passive fatigue is characterized by boredom and underload (for example, driving on a monotonous road with little traffic). Vehicle automation reduces the driver's control interactions with the vehicle. In relation to the transactional model of fatigue (Desmond and Matthews, 2009), active fatigue would appear to relate to increased task-focused coping, whereas passive fatigue seems to correspond to reduced task-focused, and increased avoidance coping. Thus, we hypothesized that active and passive fatigue manipulations would elicit differing subjective state and coping responses.

In a simulator study ($N = 108$), Saxby et al. (2008) compared the impact of three different simulator drives on stress and fatigue, manipulated between-subjects. All three drives followed the same road layout, including curves, hills, and background scenery such as bridges and gas stations, to provide some interest to the driver. In a control condition, driving was normal. In an active fatigue condition, simulated 'wind gusts' were implemented to make the vehicle more difficult to steer. In a passive fatigue condition, speed and steering were under full automation. To ensure some task involvement in this condition, the driver was asked to monitor the screen and press the turn signal on detecting an occasional 'automation failure'.

The transactional perspective emphasizes the dynamic nature of operator response to task demands; stress and fatigue states may develop (or dissipate) over time. Thus, duration of the drive (10, 30, or 50 minutes) was also manipulated in the study, between subjects. The Dundee Stress State Questionnaire (DSSQ: Matthews et al., 2002), including embedded appraisal and coping scales, and an additional fatigue questionnaire (Hitchcock and Matthews, 2005), were used to evaluate how subjective response varied across conditions. Workload was assessed using the NASA Task Load Index (NASA-TLX: Hart and Staveland, 1998).

Figure 12.1 shows state change in each condition, in terms of the three principal factors of the DSSQ. The state changes are scaled as standard scores. Consistent with expectation, the different drives elicited different patterns of subjective state. Active fatigue provoked the greatest increase in distress (> 1 SD), whereas passive fatigue produced the lowest task engagement. These state changes varied somewhat with duration, with declines in task engagement developing more rapidly in the passive fatigue than in the control condition. The active fatigue manipulation also produced some loss of task engagement after 50 minutes, as well as increased muscular fatigue, measured by the Hitchcock and Matthews (2005) instrument.

Effects of task condition on additional scales were also consistent with these state changes. For example, active fatigue elicited the highest levels of workload on the NASA-TLX, and also produced elevated levels of task-focused and emotion-focused coping. This pattern of change is consistent with a 'stressful' experience in the active fatigue condition, as the driver strives to cope with the need to counter the wind gusts, and also to manage their own distress response. Conversely, the passive fatigue condition was characterized by low workload, low challenge appraisals (consistent with the experience being monotonous), and higher levels of avoidance coping than those seen in the control condition.

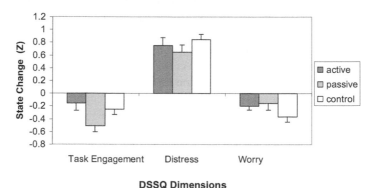

Figure 12.1   Changes in subjective state for active fatigue, passive fatigue, and control conditions in simulated driving

Thus, automation may indeed provoke a 'passive fatigue' state that elicits tired feelings, erodes task motivation and impairs concentration, the three defining features of the low task engagement state (Matthews et al., 2002). Drivers of automated vehicles may be especially vulnerable to passive fatigue, even after reverting to normal vehicle control.

In the Saxby et al. (2007) study automation was externally imposed on the driver, who had no ability to switch it on or off. This external control of automation corresponds to future intelligent highways on which automation will be mandatory. In other circumstances, as with contemporary ACC systems, the driver is free to engage or disengage automation as preferred. Possibly, greater perceived control over automation would protect against stress responses.

In a further study (N = 190: Neubauer et al., forthcoming), drivers were voluntarily able to engage in full automation for five-minute periods. The manipulation allowed a test of how voluntary choice to automate the vehicle relates to subjective fatigue. Potentially, this procedure would give the driver a rest break from the demands of driving, and alleviate fatigue. On the other hand, use of automation would likely do little to elevate the challenge of the task, or reduce monotony.

Several findings substantiated the potential of automation to produce passive fatigue. First, the availability of automation had no effect on fatigue and stress responses; declines in task engagement and increases in distress were similar, irrespective of whether the driver had the option of using automation. Within the 'automation-available' condition, even those drivers who chose to use automation showed large declines in task engagement; strategic use of automation did not protect against fatigue. Second, drivers who were initially lower in task engagement were more likely to use automation subsequently. Fatigue may encourage use of automation, even though its effect may be to increase fatigue. Third, drivers who used automation experienced increased distress. Thus, use of automation may actually tend to exacerbate driver fatigue and stress. The findings complement those of Saxby et al. (2008) in suggesting that use of vehicle automation does nothing to protect the driver against passive fatigue.

The effects of automation may be less pronounced when automation is partial, and the driver retains responsibility for some control operations. Funke et al. (2007) investigated the effects of a partial-automation manipulation in which speed control was automated but the driver was required to steer. This manipulation reduced distress compared to a drive with normal driver control of speed. Task engagement declined during the drive, suggesting the build-up of fatigue, but the loss of engagement did significantly exceed that seen in normal driving conditions. Partial automation did not reduce workload, but it was associated with improved attention to road-side stimuli, suggesting that it allowed the driver to deploy attention more effectively than in normal driving. Thus, driver control of steering, plus attentional demands of the traffic environment, appear to protect against passive fatigue.

Effects of Automated Driving on Performance and Safety

Thus far, we have argued that full vehicle automation elicits a passive fatigue state characterized by loss of task engagement, lack of challenge and avoidance coping. The general dangers of fatigue are well understood (Williamson et al., 2011). Simulator studies have identified a variety of behavioural outcomes of fatigue, including deterioration in steering of the vehicle and loss of attention to the external traffic environment (Matthews et al., forthcoming b; Philip et al., 2003). However, rather little work has been done to show that passive fatigue is more dangerous than active fatigue, as hypothesized by Desmond and Hancock (2001).

A study compatible with the passive fatigue hypothesis was reported by Matthews and Desmond (2002). They used a demanding secondary task as a fatigue induction. Workload was manipulated by having drivers follow both straight and sharply curved roadways. In fact, fatigue elicited performance deterioration was observed only under low workload conditions, which may have encouraged passive fatigue. The fatigue effect was attributed to a breakdown in effort-regulation, consistent with Hancock and Warm's (1989) dynamic model of sustained attention under stress and fatigue. The fatigued driver is motivated to reduce task-direct effort, and may believe that is safe to do so when driving conditions are perceived as routine. Furthermore, drivers in this study also showed a reduced frequency in small-magnitude steering responses, suggesting loss of task-directed effort, on straight road sections.

Saxby et al. (2008) conducted a second simulator study ($N = 168$) that aimed to compare directly the effects of active and passive fatigue on performance. The study employed the same manipulations to induce fatigue – wind gusts and automation – as Saxby et al.'s (2007) earlier study. Driving durations were 10 and 30 minutes, because Saxby et al. (2007) found that the loss of task engagement induced by passive fatigue appeared to stabilize after 30 minutes. The study featured a five-minute interval following fatigue induction, in which normal control was restored in all conditions. Performance was assessed during this interval. To test driver alertness, a van pulled out in front of the driver unexpectedly, requiring a steering and/or braking response in order to avoid collision.

Results showed the expected pattern of subjective responses to the manipulations. As in the Saxby et al. (2007) study, passive fatigue was associated with lower workload and also lower task engagement at the longer of the two drive durations. The effects of the three driving conditions on steering and braking response times are shown in Figure 12.2 (averaged across duration). Active fatigue was associated with faster response times, especially for steering, and passive fatigue produced the slowest response latencies. In addition, the fatigue manipulation elicited the highest frequency of actually crashing into the van. Averaging across duration, the crash percentages were 68 per cent (control), 54 per cent (active fatigue) and 93 per cent (passive fatigue). Indeed, at the longer, 30 minute duration, none of the 28 participants in the passive fatigue condition avoided crashing. Thus, the loss of task engagement associated with passive fatigue appears to be accompanied by loss of alertness and situation awareness.

Broadly, these results support Desmond and Hancock's (2001) conceptualization of the dangers of passive fatigue, as well as the hypothesis that passive fatigue may disrupt appropriate matching of effort to task demands (Matthews and Desmond, 2002; Young and Stanton, 2007). Drivers may become reluctant to exert effort in scanning the traffic environment for potential hazards. However, there are two issues that call for some further discussion – first, fatigue effects on steering responses, and, second, the potential for overload as well as underload effects.

In the Matthews and Desmond (2002) studies, although some influence of fatigue on alertness to roadside objects was found, the most robust finding was that fatigue impaired lateral control of the vehicle and reduced small-magnitude steering movements, on straight roads. Saxby et al. (2008) obtained a different outcome. Following the period of automation, drivers in the passive fatigue condition showed only a transient increase in positional variability, linked to having to re-establish manual control following the period of automation. Thus, the Matthews and Desmond (2002) finding was not replicated.

The difference in findings between the two studies may reflect differences in the manner of fatigue induction. In the Matthews and Desmond (2002) studies, the driver had to maintain normal vehicle control throughout, with fatigue being induced by a demanding secondary task requiring vigilance to road-signs. In the Saxby et al. (2008) study, the driver experienced monotony without having to steer. The Matthews and Desmond (2002) findings may be tied to the prolonged demand for vehicle control. Fatigue may have encouraged an aversion to this specific, repetitive task component that led to effort being withdrawn from steering on straight roads, consistent with classic studies in fatigue that show a switch in activities is often sufficient to restore performance (Holding, 1983). By contrast, the fatigued drivers in Saxby et al. (2008), although lacking alertness, may have

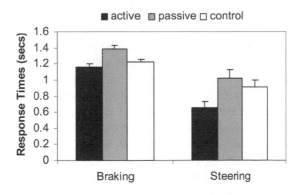

Figure 12.2   Response times for braking and steering state for active fatigue, passive fatigue, and control conditions in simulated driving

had no specific aversion to directing effort towards steering. It follows that the alertness metrics (response to emergency event) may have been more sensitive to the fatigue state than variability of lateral position.

Methodologically, the findings underscore the need for multiple performance indices in evaluating the impact of fatigue on performance. They also suggest that countermeasures based on analysis of steering movements may be ineffective in detecting loss of alertness associated with fatigue. Rumble strips that signal impairment in vehicle control may be similarly ineffective (cf., Williamson, forthcoming).

Turning to the overload issue, the studies reviewed do not address whether fatigue may also be harmful to performance when task load is high, as some real-life studies suggest (Friswell and Williamson, 2008). Although Matthews and Desmond (2002) demonstrated that fatigue was irrelevant to performance during the moderate workload of driving a series of sharp 'S-bends', workload might be substantially higher in driving environments characterized by high traffic volumes and challenging physical conditions. For example, Funke et al. (2007) showed that inducing stress through reducing the coefficient of friction of the roadway to simulate icing elevated workload. Thus, a task for future research is to investigate whether passive fatigue also leads to a depletion of attentional capacity or resources that would lead to performance impairments at high workloads.

A clue to the possibility of overload effects comes from a recent study (Reinerman et al., 2008). They used transcranial Doppler sonography (TCD) to measure cerebral blood velocity (CBFV) in the medial cerebral arteries. Previous studies (Warm et al., 2008) have shown that the vigilance decrement in performance on tasks requiring sustained attention is accompanied by declining CBFV. The change in CBFV is not simply a consequence of monotony or declining arousal. No change in CBFV is seen if the participant passively watches task stimuli, without attempting to respond. The index appears to be directly linked to cognitive workload, and, perhaps, to the depletion of resources associated with sustained attention on cognitively demanding monitoring tasks (Warm et al., 2008). Both CBFV and subjective task engagement predict performance on a range of vigilance tasks (Matthews et al., 2010).

Reinerman et al. (2008) recorded CBFV during performance of a monotonous, 36-minute drive. A large-magnitude decline in subjective task engagement was observed, together with a performance deficit, increasing SD of lateral position as the drive progressed. The TCD data showed a progressive decline in CBFV over time that was evident in both hemispheres. As in vigilance studies (Matthews et al., 2010), subjective task engagement was positively correlated with CBFV. We might interpret these results as showing a depletion of general attentional resources during the drive. However, by contrast with vigilance studies (Matthews et al., 2010), there were no significant associations between measures of CBFV response and performance indices. In underload conditions, depletion of resources may not lead to any direct performance deficit. However, should workload increase, the fatigued driver may experience inadequate resources for maintaining effective attention, for example when multiple vehicles must be monitored.

Individual Differences in Fatigue Response

Drivers differ substantially in their tolerance of prolonged driving. Matthews et al. (1997) identified five traits linked to the driving context that govern the individual's vulnerability to stress and fatigue. These are aggression, dislike of driving, fatigue proneness, hazard monitoring, and thrill seeking, as measured by the Driver Stress Inventory (DSI: Matthews et al., 1997). Within the transactional model of driver stress (Matthews, 2002; Neubauer et al., forthcoming), these traits may be related to characteristic patterns of appraisal and coping elicited by the traffic environment. For example, aggression is linked to hostile appraisals of other drivers, and dislike of driving to negative self-appraisals.

Consistent with the cognitive bases for the traits of the DSI, they are also reliably associated with appropriate emotional responses in both simulator and field studies. For example, aggression is associated with anger responses to driving, and dislike of driving relates to negative mood and distress (see Desmond and Matthews, 2009; Matthews, 2002). Fatigue proneness is the most reliable predictor of loss of task engagement in these studies; hazard monitoring relates somewhat less robustly to higher engagement (see Desmond and Matthews, 2009; Matthews, 2002). Matthews and Desmond (1998) showed in a simulator study that fatigue proneness predicts post-drive engagement even with pre-drive engagement statistically controlled. That is, fatigue proneness seems to relate to the change in fatigue state directly induced by the experience of driving. A similar result was obtained for real driving by Desmond and Matthews (2009).

Our recent studies provide some further insights into the nature of individual differences in fatigue states during automated driving. Table 12.1 gives the correlations between selected DSI scales and post-drive state in two of our recent studies (Neubauer et al., forthcoming; Saxby et al., 2008). In addition to associations between fatigue-proneness and state response, the table also demonstrates the roles of hazard monitoring in resisting fatigue, and dislike of driving as a trait that exacerbates distress and worry. The table pools data from the various experimental groups previously described, but the relationships shown were fairly constant across the different groups. However, evidence was found for influences of drive duration and use of automation on the state correlates of fatigue proneness.

In the Saxby et al. (2008; unpublished analyses) study, the relationship between fatigue-proneness and post-drive mental state varied with drive duration. After 10 minutes, fatigue-proneness related to distress ($r = .41$, $p < .01. 01$, $N = 90$), but not to task engagement ($r = -.13$). In the 30-minute condition, the corresponding correlations were .37 and -.42 ($p < .01$, $N = 90$, for both). Thus, while fatigue-proneness is associated with higher distress from an early stage, it takes longer

Table 12.1   Correlations between post-drive subjective states and selected
DSI stress vulnerability scales, in two recent studies

		DSSQ factor		
DSI scale		Engagement	Distress	Worry
Fatigue-Proneness	Saxby et al. (2008)	-.27**	.39**	24**
	Neubauer et al. (forthcoming)	-.48**	.44**	.16*
Dislike of Driving	Saxby et al. (2008)	-.06	.30**	.32**
	Neubauer et al. (forthcoming)	-.15*	.36**	.33**
Hazard Monitoring	Saxby et al. (2008)	.25**	-.16	.00
	Neubauer et al. (forthcoming)	.28**	-.17	-.04

Note: *$p < .05$, *$p < .01$.

for fatigue-prone drivers to show loss of task engagement. Similarly, fatigue-proneness was unrelated to avoidance coping after 10 minutes ($r = .18$), but the correlation with avoidance reached significance after 20 minutes ($r = .24$, $p < .05$). Perhaps use of avoidance to deal with distress leads to later loss of task engagement.

Neubauer et al. (forthcoming) found that fatigue proneness tended to be associated with lower task engagement and higher distress in both non-automated and automation-optional conditions. However, when drivers were given the option of using automation, fatigue-proneness was more strongly associated with distress in automation users than in drivers who chose to maintain manual control. Neubauer et al. (forthcoming) suggest that automation increases awareness of the discomforts of the fatigue state, which in turn elevates distress. More generally, the study shows that automation does nothing to mitigate fatigue in drivers vulnerable to the state.

Conclusions

The studies reviewed have several implications. First, they provide further support to those authors who have identified possible dangers as well as benefits to vehicle automation (for example, Young and Stanton, 2007). Consistent with Desmond and Hancock's (2001) account of passive fatigue, full vehicle automation tends to induce a passive fatigue state, even though it lowers workload. In relation to cognitive models of stress, lack of challenge and use of avoidance coping appear to promote passive fatigue. The subjective expressions of passive fatigue, notably low task engagement, are accompanied by loss of alertness, evidenced by slowed responses to an emergency event. Allowing the driver to initiate automation voluntarily provides no protection against passive fatigue.

The findings have several practical applications for development of in-vehicle automation. Generally, it is important for system designers to evaluate vehicle automation effects on fatigue and alertness during driving. There may be design solutions that increase the driver's active involvement with the task to counteract fatigue. In addition, partial automation seems less prone to produce fatigue (Funke et al., 2007). The results also emphasize that fatigue countermeasures need to be directed not just towards sleep deficits, but also towards the disengaged, but wakeful driver.

The relationships found between the DSI fatigue proneness scale and subjective fatigue and distress, suggest that questionnaire measures may be used to identify drivers who are especially fatigue-prone. Such measures may be used in selection of professional drivers at risk from passive fatigue, such as long-haul truck drivers. Associations between fatigue-proneness and cognitive stress processes also support the utility of interventions based on enhancing driver coping (Dorn et al., 2010; Machin and Hoare, 2008). Drivers who are generally fatigue-prone are not immune to the adverse effects of vehicle automation. Indeed, fatigue-prone drivers may be especially vulnerable to distress during automated driving.

References

Brown, I.D. (1997). Prospects for technological countermeasures against driver fatigue. *Accident Analysis and Prevention, 29*, 525–31.

Desmond, P.A., and Hancock, P.A. (2001). Active and passive fatigue states. In P.A. Hancock and P.A. Desmond (Eds.), *Stress, workload, and fatigue* (pp. 455–65). Mahwah, NJ: Lawrence Erlbaum.

Desmond, P.A., and Matthews, G. (2009). Individual differences in stress and fatigue in two field studies of driving. *Transportation Research Part F: Traffic Psychology and Behaviour, 12*, 265–76.

Dorn, L., Stephen, L., af Wåhlberg, A., and Gandolfi, J. (2010). Development and validation of a self-report measure of bus driver behaviour. *Ergonomics, 53*, 1420–33.

Fairclough, S.H. (2001). Mental effort regulation and the functional impairment of the driver. In P.A. Hancock and P.A. Desmond (Eds.), *Stress, Workload, and Fatigue*. Mahwah, NJ: Lawrence Erlbaum.

Friswell, R., and Williamson, A. (2008). Exploratory study of fatigue in light and short haul transport drivers in NSW, Australia. *Accident Analysis and Prevention, 40*, 410–17.

Funke, G.J., Matthews, G., Warm, J.S., and Emo, A.K. (2007). Vehicle automation: A remedy for driver stress? *Ergonomics, 50*, 1302–23.

Hancock, P.A., and Warm, J.S. (1989). A dynamic model of stress and sustained attention. *Human Factors, 31*, 519–37.

Hart, S.G., and Staveland, L.E. (1988). Development of NASA-TLX (Task Load Index): Results of empirical and theoretical research. In P.A. Hancock and N. Meshkati (Eds.), *Human Mental Workload* (pp. 139–83). Amsterdam: Elsevier.

Hitchcock, E.M., and Matthews, G. (2005). Multidimensional assessment of fatigue: A review and recommendations. *Proceedings of the International Conference on Fatigue Management in Transportation Operations*, Seattle, WA, September 2005.

Hockey, G.R.J. (2011). A motivational control theory of cognitive fatigue. In P.L. Ackerman (Ed.), *Cognitive Fatigue: Multidisciplinary perspectives on current research and future applications* (pp. 167–87). Washington, DC: American Psychological Association.

Holding, D.H. (1983). Fatigue. In G.R.J. Hockey (Ed.), *Stress and Fatigue in Human Performance* (pp. 145–68). Chichester, UK: Wiley.

Lazarus, R.S. (1999). *Stress and emotion: A new synthesis*. New York: Springer.

Machin, M. A., and Hoare, P. N. (2008). The role of workload and driver coping styles in predicting bus drivers' need for recovery, positive and negative affect, and physical symptoms. *Anxiety, Stress & Coping: An International Journal, 21*, 359–375.

Matthews, G. (2002). Towards a transactional ergonomics for driver stress and fatigue. *Theoretical Issues in Ergonomics Science, 3*, 195–211.

Matthews, G., Campbell, S.E., Falconer, S., Joyner, L., Huggins, J., Gilliland, K., Grier, R., and Warm, J.S. (2002). Fundamental dimensions of subjective state in performance settings: Task engagement, distress and worry. *Emotion, 2*, 315–40.

Matthews, G., Davies, D.R., Westerman, S.J., and Stammers, R.B. (2000). *Human Performance: Cognition, stress and individual differences*. London: Psychology Press.

Matthews, G., and Desmond, P.A. (1998). Personality and multiple dimensions of task-induced fatigue: A study of simulated driving. *Personality and Individual Differences, 25*, 443–58.

Matthews, G., and Desmond, P.A. (2002). Task-induced fatigue states and simulated driving performance. *Quarterly Journal of Experimental Psychology: Human Experimental Psychology, 55*, 659–86.

Matthews, G., Desmond, P.A., and Hitchcock, E.M. (forthcoming a). Dimensional models of fatigue. In P.A. Desmond, G. Matthews, P.A. Hancock, and C.E. Neubauer (Eds.), *Handbook of Operator Fatigue*. Aldershot, UK: Ashgate.

Matthews, G., Desmond, P.A., Joyner, L.A., and Carcary, B. (1997). A comprehensive questionnaire measure of driver stress and affect. In E. Carbonell Vaya and J.A. Rothengatter (Eds.), *Traffic and Transport Psychology: Theory and Application* (pp. 317–24). Amsterdam: Pergamon.

Matthews, G., Saxby, D.J., Funke, G.J., Emo, A.K., and Desmond, P.A. (forthcoming b). Driving in states of fatigue or stress. In D. Fisher, M. Rizzo,

J. Caird and J. Lee (Eds.), *Handbook of Driving Simulation for Engineering, Medicine and Psychology.* Boca Raton, FL: Taylor and Francis.

Matthews, G., Warm, J.S., Reinerman-Jones, L.E., Langheim, L.K., Washburn, D.A., and Tripp, L. (2010). Task engagement, cerebral blood flow velocity, and diagnostic monitoring for sustained attention. *Journal of Experimental Psychology. Applied, 16,* 187–203.

Neubauer, C., Langheim, L., Matthews, G., and Saxby, D. (forthcoming a). Fatigue and voluntary utilization of automation in simulated driving. *Human Factors.*

Neubauer, C., Matthews, G., and Saxby, D. (forthcoming b). Driver fatigue and safety. In P.A. Desmond, G. Matthews, P.A. Hancock and C.E. Neubauer (Eds.), *Handbook of Operator Fatigue.* Aldershot, UK: Ashgate.

Philip, P., Sagaspe, P., Taillard, J., Moore, N., Guilleminault, C., Sanchez-Ortuno, M., Akerstedt, T., and Bioulac, B. (2003). Fatigue, sleep restriction, and performance in automobile drivers: A controlled study in a natural environment. *Fatigue, Sleep Restriction, and Performance, 26,* 277–80.

Reinerman, L.E., Warm, J.S., Matthews, G., and Langheim, L.K. (2008). Cerebral blood flow velocity and subjective state as indices of resource utilization during sustained driving. *Proceedings of the Human Factors and Ergonomics Society, 52,* 1252–56.

Saxby, D.J., Matthews, G., Hitchcock, E.M., and Warm, J.S. (2007). Development of active and passive fatigue manipulations using a driving simulator. *Proceedings of the Human Factors and Ergonomics Society, 51,* 1237–41.

Saxby, D.J., Matthews, G., Hitchcock, E.M., Warm, J.S., Funke, G.J., and Gantzer, T. (2008). Effects of active and passive fatigue on performance using a driving simulator. *Proceedings of the Human Factors and Ergonomics Society 52,* 1252–56.

Stanton, N., and Young, M.S. (2005). Driver behaviour with adaptive cruise control. *Ergonomics, 48,* 1294–1313.

Warm, J.S., Parasuraman, R., and Matthews, G. (2008). Vigilance requires hard mental workload and is stressful. *Human Factors, 50,* 433–41.

Williamson, A. (forthcoming). Countermeasures for driver fatigue. In P.A. Desmond, G. Matthews, P.A. Hancock and C.E. Neubauer (Eds), *Handbook of Operator Fatigue.* Aldershot, UK: Ashgate.

Williamson, A., Lombardi, D.A., Folkard, S., Stutts, J., Courtney, T.K., and Connor, J.L. (2011). The link between fatigue and safety. *Accident Analysis and Prevention, 43,* 498–515.

Young, M.S., and Stanton, A. (2007). Back to the future: Brake reaction times for manual and automated vehicles. *Ergonomics, 50,* 46–58.

Chapter 13

Knowledge of Traffic Hazards: Does it Make a Difference for Safety?

Anders af Wåhlberg
Department of Psychology, Uppsala University, Sweden

Lisa Dorn
*Department of Systems Engineering and Human Factors,
Cranfield University, UK*

Introduction

Regulatory authorities and organizations keen to improve road safety often use knowledge of rules and laws of traffic as a measure of driving safety (see for example the procedure in Janke, 1990, where older drivers had to pass a test of traffic law to be able to renew their licences). Over time, such test content has also been supplemented by questions about vehicle handling (for example stopping distances) and the hazards involved (such as skidding in wet weather). Yet, knowledge of rules, laws and hazards seem to have little bearing on actual safety. Thus, in Janke (1990), the (randomly allocated) drivers who had to pass a law test had subsequent driving records that were no different in crash involvement to those who did not take part in such a test.

What little literature exists in this field seems to be in fair agreement with the Janke study. Arthur and Doverspike (2001) found a correlation of -.19 between a knowledge test modelled from a US state department driving manual and self-reported at-fault crashes over three years, although it is not known whether these crashes were reported after testing. A similarly weak association (correlation about .21) between driving knowledge and recorded crashes was reported by Barkley et al. (2002) in a mixed sample of young drivers diagnosed with Attention Deficit Hyperactivity Disorder and controls.

Conley and Smiley (1976) used a different kind of outcome variable; driving violations. In a sample of more than 20,000 drivers, they found no associations of any significance between errors on a driving knowledge test and traffic violations over a period of more than four years. They concluded that passing this type of test was a mere formality which had nothing to do with safety.

Gebers (1995) correlated knowledge acquisition on a course for traffic violators with their accidents and violations. Only the latter was significantly associated with an increase in knowledge (-.09). In general, the evidence to date would therefore seem to indicate a very weak association, at best, between knowledge of driving rules and safety.

However, a standard argument used against such results is that studies have low statistical power when low variance variables like crashes and offences are used as outcomes, which make it improbable to find an effect. Instead, it is often claimed that self-reported driver behaviour can be a substitute dependent measure of traffic risk (for a review of this line of argument, see af Wåhlberg, 2009).

It could also be argued that even though driving knowledge may have extremely little impact on safety, it should co-vary with behaviours that are often canvassed in driver inventories, because a higher level of knowledge of risks should make drivers refrain from such behaviours as aggression and aberrant driving behaviours, even though these might not actually be related to safety.

The aim of the present study was therefore to test whether knowledge of driving risks was associated with self-reported risky behaviours and outcomes such as self-reported crash involvement and driving offences.

Method

General

The present study was undertaken as part of an evaluation of a driver improvement scheme run by Thames Valley Police in the UK. Here, drivers choose to attend a traffic safety course instead of having penalty points added to their driving records, and instead of paying a fine, participants paid a similar fee to take part in the course. The course was only available for drivers under the age of 25 and called the Young Driver Scheme (YDS). The YDS content consisted of an initial workshop session covering a variety of topics on driving safety, such as: peer pressure, use of drugs and alcohol, stopping distances and hazard awareness (for details, see af Wåhlberg, 2010). The rest of the course was distributed online over the web in five different modules which took about 30 minutes to complete, and covered similar subjects as covered in the workshop, but in more detail. Four of these modules are completed when the participants go through an assessment consisting of 25 questions on the content of the module. To pass each module and progress through the course to the next module, the driver is required to achieve 20 out of 25 questions correct.

As part of a study to evaluate the effectiveness of the YDS, a questionnaire was also distributed online before and after the online modules. Police-recorded penalty-points data from the national driver records database was also collected for some drivers. Due to response attrition and differing numbers being available at different times, the number of drivers included in different analyses varied substantially.

Samples

The project involved a sample of young drivers (< 25 years) who took part in YDS course. All drivers completed the online questionnaire before the start of the online modules and about 80 per cent completed the questionnaire at the end of the course. About 20 per cent of the drivers completed the questionnaire again six months later. The data set therefore contained unequal numbers of drivers for the different questionnaire waves.

Variables

The scores from the knowledge assessment for each of the four YDS modules were used as a measure of reportable knowledge of driving rules and hazards. As the driver was required to pass each of the four assessments, there were several possible knowledge variables, including: Failure, Safety Margins, Overtaking, Attitude and Alertness, Anticipation and Hazard Perception.

The number of correct responses when a driver passed each of the four assessments individually formed the first variable. As there were four assessments, one variable for the mean level of correct responses across all assessments was also computed. Finally, the number of failed assessments for each driver formed another variable. Therefore, six dependent measures were used for the present analyses.

Knowledge of driving risks and rules was tested as the current number of correct responses to questions about course content so the results can be viewed in two ways. They could be indicative of the level of knowledge after the course, but also to some degree as being dependent upon what knowledge the driver brought along to the course. The assumption here is that those with greater initial driving knowledge would achieve better scores on the assessments.

Therefore, several measurements could be viewed as dependent variables, depending on when the measurement took place. Thus, each wave of questionnaire administration could be associated with the knowledge variable. Variables to be used were therefore all the driver inventories administrated on three occasions, plus self-reported crashes and penalty points in waves 1 and 3.

Four well-known driver inventories were included in the online questionnaire: the Driving Anger Scale (DAS; Deffenbacher et al., 1994), the violation subscale of the Manchester Driver Behaviour Questionnaire (DBQ; Reason et al., 1990), the two-item version of the Sensation Seeking Scale (SSS; Slater, 2003), and the Aggression scale from the Driver Behaviour Inventory (DBI), a driver stress measure (Gulian et al., 1988). Three items on the use of alcohol and drugs in relation to driving were also added as a separate scale. All these scales have been reported to predict self-reported traffic accidents (Beirness and Simpson, 1988; Davey et al., 2007; Deffenbacher et al., 2004; Horwood and Fergusson, 2000; King and Parker, 2008; Knee et al., 2001; Matthews et al., 1999; Matthews et al., 1991; Matthews et al. 1997).

144   *Advances in Traffic Psychology*

Next, items measuring self-reported driving experience, mileage per month, total number of collisions and current number of licence penalty points were also included, before and after the course. It should be noted that the first collision variable asked specifically for all crashes since licensing, making no effort to control for exposure in any way (meaning that there was no set period of time over which crashes were reported). Exposure was instead controlled for by using the experience and mileage variables.

The number of recorded penalty points for some of the participants was also available. It should be noted that this variable was not the same as the number of self-reported penalty points.

Results

Descriptive data for the respondents in the first questionnaire wave can be viewed in Table 13.1. It should be noted that this sample was almost 100 per cent of the drivers taking the course, as they were required to respond to the questionnaire before starting the course.

First, the assessment scores and the failure variable were correlated. The latter tended to be negatively associated with the individual assessment scores, with -.11 (the Overtaking module) being the strongest correlation ($N = 4027$). The assessment scores correlated from -.01 to .22 between each other.

The scores on each of the four assessments (when passed) as well as the mean of these, and number of failed assessments were correlated with the driver inventory scales, self-reported crashes and total penalty points, penalty points per mile (experience multiplied by miles per month), age and experience ($N = 3925$) taken from the first wave of the questionnaire. The strongest correlation found was between the score on the overtaking module and experience (years of being licenced), which was .11. Removing some outliers from the crashes and points per mile variables did not alter these associations in any significant way.

The driver inventory scale scores from the second wave questionnaire were also correlated with the assessment scores and failures. The strongest association was .08, for sensation seeking and the score for the overtaking assessment ($N = 3199$). It should be noted that this correlation was positive, meaning that those with higher knowledge reported higher levels of sensation seeking.

Table 13.1   Descriptive data for the first wave respondents in the study. Shown are the percentage of men, mean and standard deviation of age, time since licensing, number of self-reported crashes since licensing, and current number of penalty points on licence

Wave	N	Sex	Age	Experience	Crashes	Points
1	4023	59.0%	21.7/2.2	3.2/2.1	0.578/0.867	0.704/1.635

For the last wave of the questionnaire, a slightly different result emerged. These correlations are therefore reported in Table 13.2. It can be seen that self-reported number of penalty points after the course appears to have some negative association with driving knowledge, which would seem to be a reasonable result. However, the strongest correlations were for the sensation seeking scale, and again, these were positive.

The highest correlation between the assessment scores and the total number of penalty points, recorded before the course by the Driver and Vehicle Licensing Agency, was .08 ($N = 524$). The correlations for points recorded after the course and assessment scores were lower.

Table 13.2   The correlations between wave three items and scales and the assessment variables. Collisions and penalty points were reported for the previous six months ($N = 318$)

Variable	Collisions	Points	DAS	DBI-A	SSS	Drugs	DBQ-V
Failures	-.03	-.01	-.02	-.02	-.07	-.05	-.02
Safety margins	-.07	-.11*	-.02	.00	.18***	.06	.11
Overtaking	-.05	-.07	.10	.03	.17**	.05	.12*
Attitude and alertness	.01	-.02	-.07	-.03	-.04	-.03	-.04
Anticipation and hazard perception	.00	-.11*	.04	-.03	.08	-.02	-.03
All modules mean	-.04	-.14*	.01	-.02	.16**	.02	.06

Note: * $p < .05$, ** $p < .01$, *** $p < .001$

Discussion

The present study showed no associations of any practical significance between reportable knowledge of traffic hazards and any of a number of possible indicators of dangerous driving behaviour and/or crash risk. There would therefore seem to be little point in using driving knowledge as an indicator of driving safety (see Eby et al., 2003).

It can be noted that the assessment scores tended to be very weakly correlated between themselves and with the number of failures. This would seem to indicate that there is little stability in knowledge (or stability is limited to very specific sub-areas), regardless of whether the present results are seen as indicative of the level of knowledge the drivers brought with them, or of what they learned. Instead, their scores would seem to be rather random.

The strongest associations found between the assessment scores and any other variable was for sensation seeking. However, these tended to be positive, which would seem to reflect poorly on either the knowledge test or on the sensation seeking inventory, or both. They cannot both be indicators of safety, as claimed. Similar conclusions can be drawn from the extremely weak correlations for the other inventories.

It could, of course, still be argued that a basic road safety knowledge is necessary for safe driving, and that the lack of effects shown so far are due to ceiling effects given the high levels of knowledge in the drivers tested. This kind of argument would, however, limit the use of knowledge tests to the very early stages of driving, in other words learner driving, while for licensing purposes, it may not be valid. If any such use is contemplated, it would need to be supported by some more positive evidence than that available to date.

The present results may to some degree be in disagreement with those of other researchers. For example, Legree et al. (2003) found that a test of knowledge of safe speed and accident causation had some associations with crashes. However, as the differences between that study and the present one were many, this discrepancy may only be superficial. For example, Legree et al. used an amalgamated criterion of recorded and self-reported crashes, and a very unusual scoring method for the driver responses, none of which were utilized in the present study. Furthermore, the strongest correlation observed was -.19, in other words 3.6 per cent of the explained variance, and small chance deviations may therefore explain the differences between our respective studies. Similarly weak effects and similar methodological differences can also be noted for Arthur and Doverspike (2001) and Barkley et al. (2002).

The main problem, however, may be the heterogeneity of the knowledge tests used in different studies, which may explain the differences. However, such a claim would also undermine the use of knowledge tests in general, because it would point out the need to validate each specific test (used by licensing authorities for example). Today, hardly any such validations appear to have been undertaken, although it is possible that driving authorities have undertaken such tests without publishing them. Without available evidence, however, the use of knowledge tests within traffic safety can be seriously questioned.

References

Arthur, W. Jr., and Doverspike, D. (2001). Predicting motor vehicle crash involvement from a personality measure and a driving knowledge test. *Journal of Prevention and Intervention in the Community, 22*, 35–42.

Barkley, R.A., Murphy, K.R., Du Paul, G.J., and Bush, T. (2002). Driving in young adults with attention deficit hyperactivity disorder: Knowledge, performance, adverse outcomes, and the role of executive functioning. *Journal of the International Neuropsychological Society, 8*, 655–72.

Beirness, D.J., and Simpson, H.M. (1988). Lifestyle correlates of risky driving and accident involvement among youth. *Alcohol, Drugs and Driving, 4*, 193–204.

Conley, J.A., and Smiley, R. (1976). Driver licensing tests as a predictor of subsequent violations. *Human Factors, 18*, 565–74.

Davey, J., Wishart, D., Freeman, J., and Watson, B. (2007). An application of the driver behaviour questionnaire in an Australian organizational fleet setting. *Transportation Research Part F, 10*, 11–21.

Deffenbacher, J.L., Oetting, E.R., and Lynch, R.S. (1994). Development of a driving anger scale. *Psychological Reports, 74*, 83–91.

Deffenbacher, J.L., White, G.S., and Lynch, R.S. (2004). Evaluation of two new scales assessing driver anger: The Driver Anger Expression Inventory and the Driver's Angry Thoughts Questionnaire. *Journal of Psychopathology and Behavioral Assessment, 26*, 87–99.

Eby, D.W., Molnar, L.J., Shope, J.T., Vivoda, J.M., and Fordyce, T.A. (2003). Improving older driver knowledge and self-awareness through self-assessment: The driving decisions workbook. *Journal of Safety Research, 34*, 371–81.

Gebers, M.A. (1995). *Knowledge and Attitude Change and the Relationship to Driving Performance among Drivers Attending California Traffic Violator School*. RSS-95-147. California Department of Motor Vehicles.

Gulian, E., Glendon, A.I., Matthews, G., Davies, D.R., and Debney, L.M. (1988). Exploration of driver stress using self-reported data. In J.A. Rothengatter and R.A. de Bruin (Eds.), *Road User Behaviour: Theory and Research* (pp. 342–7). Maastricht: van Gorcum.

Horwood, L.J., and Fergusson, D.M. (2000). Drink driving and traffic accidents in young people. *Accident Analysis and Prevention, 32*, 805–14.

Janke, M.K. (1990). Safety effects of relaxing California's clean-record requirement for driver license renewal by mail. *Accident Analysis and Prevention, 22*, 335–49.

King, Y., and Parker, D. (2008). Driving violations, aggression and perceived consensus. *Revue Européenne de Psychologie Appliquée, 58*, 43–9.

Knee, C.R., Neighbors, C., and Vietor, N.A. (2001). Self-determination theory as a framework for understanding road rage. *Journal of Applied Social Psychology, 31*, 889–904.

Legree, P.J., Heffner, T S., Psotka, J., Martin, D.E., and Medsker, G.J. (2003). Traffic crash involvement: Experiential driving knowledge and stressful contextual antecedents. *Journal of Applied Psychology, 88*, 15–26.

Matthews, G., Desmond, P.A., Joyner, L., Carcary, B., and Gilliland, K. (1997). A comprehensive questionnaire measure of driver stress and affect. In T. Rothengatter and E.C. Vaya (Eds.), *Traffic and Transport Psychology: Theory and Application* (pp. 317–24). Amsterdam: Pergamon.

Matthews, G., Tsuda, A., Xin, G., and Ozeki, Y. (1999). Individual differences in driver stress vulnerability in a Japanese sample. *Ergonomics, 42*, 401–15.

Matthews, G., Dorn, L., and Glendon, A.I. (1991). Personality correlates of driver stress. *Personality and Individual Differences, 12*, 535–49.

Reason, J., Manstead, A., Stradling, S., Baxter, J., and Campbell, K. (1990). Errors and violations on the roads: A real distinction? *Ergonomics, 33*, 1315–32.

Slater, M.D. (2003). Alienation, aggression, and sensation-seeking as predictors of adolescent use of violent film, computer and website content. *Journal of Community, 53*, 105–21.

af Wåhlberg, A.E. (2009). *Driver Behaviour and Accident Research Methodology: Unresolved Problems*. Farnham: Ashgate.

af Wåhlberg, A.E. (2010). Re-education of young driving offenders: Effects on self-reports of driver behavior. *Journal of Safety Research, 41*, 331–8.

Chapter 14

The Theory of Planned Behaviour (TPB) and Speeding Behaviour of Young Drivers

Catherine Ferguson, Lynne Cohen, Julie Ann Pooley and
Andrew Guilfoyle
Edith Cowan University, Joondalup, Western Australia

Introduction

Those involved in road safety research will be familiar with the statistics in relation to young driver crash involvement, the range of variables that are involved, and the risky driving behaviours that are undertaken. This chapter outlines the use of the Theory of Planned Behaviour (TPB) to predict and understand the speeding behaviour of young drivers in a rural city in the South West of Western Australia.

Excess and inappropriate speed is often implicated in crashes (Clarke et al., 2006). Younger drivers and specifically young male drivers are known to drive faster than older drivers (Hatfield and Job, 2006) and male drivers have a greater affinity with speed than females (Carcary, 2002; Ferguson, 2005). Internationally, large numbers of drivers across all age groups report exceeding the speed limit (DETR, 2000: 69 per cent [UK]; Mitchell-Taverner et al., 2003: 62 per cent [Australia]).

Recent Australian research reported that approximately half of young drivers have been caught speeding between 11–25 km/h above the posted speed limit. Additionally, this longitudinal research reported reasonable consistency in speeding behaviour between participants when aged 19–20 years and again at age 23–24 years (Vassallo et al., 2010). Sixty per cent of young drivers aged 23–24 years confirmed involvement in a crash while driving, although the majority of these crashes did not involve serious injury (Vassallo et al., 2010). New Zealand research reports that 83 per cent of young drivers exceeded the speed limit by at least 20 km (Fergusson et al., 2003).

Theory of Planned Behaviour

The Theory of Planned Behaviour (TPB) is a well-known and used theory for the prediction and understanding of behaviour and has been used in a variety of studies on driver behaviour, including speeding (Conner et al., 2007; Elliott,

2010). The theory postulates that intention is the causal precursor of behaviour and that intention is predicted by attitudes toward the behaviour, subjective norms which involve the perceptions of the individual about what other people important to them think they should do regarding the behaviour, and perceived behavioural control which has been likened to self-efficacy. TPB research generally provides good results for the prediction and understanding of intention but is less successful in predicting behaviour (Armitage and Conner, 2001; Sutton, 1998). This weakness of the intention-behaviour gap in the TPB is receiving attention to improve the theory (Chatzisarantis and Hagger, 2007; Sheeran, 2002).

Unlike much of TPB research, the research described in this chapter includes the prediction of both intention and behaviour; and directs the reader to instructions for the development of a TPB questionnaire, including the elicitation of underlying beliefs. The model of the core TPB is outlined in Figure 14.1, and demonstrates the underpinning beliefs, the core TPB variables (attitude, subjective norm, and perceived behavioural control [PBC]), behavioural intention, and behaviour (for extensive reviews on the TPB and how to maximize the use of the TPB for research and interventions (see Ajzen, 1991; Ajzen 2006, Ajzen and Fishbein, 2005; Armitage and Conner 2001).

A range of additional variables have been added to the core TPB variables to convert the general theory to one that is context specific (Ajzen, 1991). The research reported in this chapter investigated both the core TPB and an 'enhanced model' that included additional variables. An important practical and theoretical issue also demonstrated in the research involved the investigation of both intention and expectation.

Although the TPB proposes that intention is the immediate precursor to behaviour, previous research, including road safety research, has often asked about expectation rather than intention. When doing so researchers have not commented on the theoretical implications. Theoretically, intention includes a motivational aspect and is required for causal explanation (Sheeran, 2002). Expectation 'externalises' the behaviour and is often used where a negative behaviour is being investigated as respondents appear more comfortable admitting to 'expecting' rather than 'intending' to undertake a particular behaviour (Gibbons et al., 1998).

Summary of the Research Reported in this Chapter

The researchers investigated both the core TPB (attitudes, subjective norms, and PBC) and an 'enhanced' model with the addition of moral norm, self-identity as a safe driver, and a driving style questionnaire from Deery and Love (1996). Young drivers were surveyed about both their intention and expectation of speeding in excess of the posted speed limit in a context that involved travelling to a social event. The questionnaire included behavioural beliefs, normative beliefs, and control beliefs that had been elicited in a previous study and tested in a pilot study and was developed based on the instructions of Ajzen (2006).

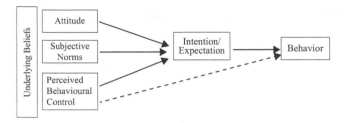

Figure 14.1   Theory of Planned Behaviour (adapted from the literature)

Participants were young adults in vocational education and training in the South West of Western Australia (n = 73). They were chosen as vocational training is a major aspect of young people's lives and the demographics of this group are closer to the general population than those of university students. The data were collected during class attendance at a local college, or training facility. This methodology facilitated the prospective design of the research which included data collection of the TPB variables up to intention and expectation at time one; and the collection of behavioural data three weeks later. The time span of three weeks allowed respondents to forget their specific responses to the first set of questions, thereby reducing demand characteristics.

The research results are presented in two parts. First the results of the TPB analyses for the prediction of speeding intention, expectation, and behaviour will be presented. Second, an analysis of the differing underpinning beliefs of those who intended to speed and those who did not intend to speed is offered.

Both logistic and standard regressions were used to analyse the data for the prediction of speeding intention, expectation, and behaviour. The measures for behaviour and behavioural intention were each dichotomised into 'any' or 'nil' as the original data produced non-normal distributions. The appropriate analysis for a dichotomised criterion variable is logistic regression and for intention two logistic regressions were used to analyse the data; the first including only the core TPB variables (attitude, SN and PBC) and the second the 'enhanced' model that included the core TPB variables plus the additional variables: the Deery and Love (1996) driving style questionnaire, questions devised for this research representing moral norm, and self-identity as a safe driver. Behavioural expectation was measured on a Likert scale range 1–7, and produced a sufficiently normal distribution to allow the use of standard multiple regression to analyse the data. Again both the core and 'enhanced' model were investigated. The intention-behaviour relationship and the expectation-behaviour relationships were explored using logistic regression.

The results of the regression analyses are shown in Table 14.1. The accounted for variance in intention and expectation differed considerably. The prediction of behaviour from each of intention and expectation was similar but at the low end of the expected range, as suggested by Sutton (1998). However, effect sizes for these results based on Cohen (1992) indicate medium to large effect sizes. Cohen suggested .02 represented a small effect size, .15 medium, and .35 large for R^2 results.

For intention, the enhanced model provided an extra 12 per cent of the accounted for variance. The Nagelkerke R^2 is similar to that of Pearson's r^2 (Nagelkerke, 1991).

There were no significant variables for the core TPB model for intention; however, the enhanced model revealed significant variables in two control questions: 'For me to drive faster than the speed limit when travelling to a social event is (easy/difficult)' and 'The decision to drive faster than the speed limit when travelling to a social event is entirely up to me'; and the Deery and Love (1996) questionnaire. The analyses of expectation revealed attitude as a significant predictor for the core TPB model; and attitude and self-identity as significant predictors for the enhanced model.

Table 14.1   Results of accounted for variance in the regression analyses

Model	Intention accounted for variance (Nagelkerke R^2)	Expectation accounted for variance R^2	Behaviour accounted for variance (Nagelkerke R^2)
Core TPB model	.21	.46	
Enhanced model	.33	.50	
Intention			.17
Expectation			.19

Comments Regarding the Prediction of Speeding Intention, Expectation and Behaviour

The accounted for variance in intention was low, but reasonable for expectation (Sutton, 1998). The intention-behaviour and expectation-behaviour relationships appeared weak, but there is extensive research currently being undertaken in addressing this gap across a variety of domains (Chatzisarantis and Hagger, 2007; Sheeran, 2002). The enhanced TPB improved the accounted for variance in intention above that of the core TPB by 12 per cent; however, both models produced similar results for expectation. The increased difference between the variance accounted for in intention compared to expectation may be a statistical issue related to measurement or may be explained by a lack of intention to speed. The theoretical differences between intention and expectation, as indicated previously, translates into practical issues in relation to how a behaviour can be modified and provide some basis for interventions. From a theoretical perspective the low variance accounted for in speeding intention suggests that the core TPB model requires improvement to ensure it is more effective for a population of young drivers. It may be suggested that there are important predictors missing from the TPB.

Analyses of Underpinning Beliefs

Six behavioural and eleven self-efficacy beliefs assessing ease and difficulty of undertaking the behaviour were collected during an elicitation study, tested in a pilot study and included in the questionnaire. Although each belief comprises a likelihood and evaluative component which is often added or multiplied to provide a composite belief (Ajzen and Fishbein, 1980), this research investigated the likelihoods and evaluations separately. This decision was made to assess the beliefs' usefulness for the development of interventions to reduce speeding behaviour in young drivers.

Two groups, non-intenders (n = 29) and intenders (n = 44), were established by dichotomising intention and a between groups *t*-test revealed significant differences for four behavioural beliefs, as shown in Table 14.2; and eight self-efficacy ease beliefs and four difficulty beliefs, as shown in Table 14.3.

All the significantly different behavioural beliefs are 'likelihood' beliefs rather than 'evaluative' beliefs and three specifically relate to crashing with perhaps less regard for the evaluation of the outcomes of the crash. These range from completely destroying their car to killing someone. The effect sizes between intenders and non-intenders indicate medium to large differences for these beliefs (Cohen, 1992).

Table 14.2   Descriptive data for behavioural beliefs significantly different 'nil' and 'any' intention (all questions are based on likelihood of (1) very likely to (7) very unlikely)

Behavioural belief	*t* test result	'nil' mean	SD	'any' mean	SD	Effect size
Driving faster than the speed limit will make me more alert on long journeys	(df1, 71) = 2.275, *p* = .026	5.14	1.70	4.14	1.92	.55sd
If I drive faster than the speed limit I will crash my car and ...						
write my car off	(df1, 71) = -2.290, *p* = .025	3.62	1.68	4.62	1.90	.56sd
injure myself or someone else	(df1, 71) = -3.284, *p* = .002	2.83	1.71	4.16	1.68	.78sd
... kill myself or someone else	(df1, 71) = -4.390, *p* = .000	2.72	1.79	4.59	1.76	1.05sd

Table 14.3   Descriptive data for self-efficacy beliefs significantly different 'nil' and 'any' intention

	'nil' mean	SD	'any' mean	SD	Effect size
Self-Efficacy – Ease Beliefs – 'It is easy for me to speed if …'					
Measured on a scale of 1–7 very true to very false					
there are no police around	3.97	2.03	2.34	1.29	.98sd
Measured on a scale of 1–7 very likely to very unlikely					
I am driving on a dual carriageway or freeway	4.79	1.57	3.48	1.73	.79sd
I am driving when the roads are quiet	4.31	1.97	3.02	1.61	.70sd
I am running late	3.28	1.69	2.50	1.41	.50sd
I am in a good mood	5.21	1.49	4.27	1.90	.55sd
I am in a bad mood	4.65	1.80	3.27	1.66	.80sd
there are no police around	4.72	1.85	3.29	1.76	.79sd
I am driving a powerful car	4.90	1.47	3.39	1.95	.88sd
Self-Efficacy – Difficult Beliefs 'It is difficult for me to speed if …'					
Measured on a scale of 1–7 very true to very false					
the weather conditions are poor	2.14	1.53	3.04	2.20	.48sd
Measured on a scale of 1–7 very likely to very unlikely					
there are a lot of police around	6.65	.61	6.18	1.22	.52sd
I am driving in heavy traffic	6.03	1.32	4.79	1.90	.77sd
I am transporting my parents	5.76	1.64	4.82	1.93	.53sd

Comments on Beliefs about Speeding

These results indicate medium to strong differences between intenders and non-intenders and raise two issues concerning the behavioural beliefs. First there is an inhibitory effect for crashing with various outcomes and second there is a facilitating effect of driving faster on long journeys to retain alertness. The self-efficacy beliefs have revealed contexts and likelihoods of speeding by young drivers which may be useful as a part of media campaigns. Specifically the perception of a police presence could be useful as there are clear differences between intenders and non-intenders on this dimension. Perhaps as part of enforcement, each police vehicle should be equipped with speed cameras. Non-intenders are inhibited by heavy traffic, transporting parents, and poor weather conditions. However, the difference for police presence between non-intenders and intenders was smaller and the mean score high suggesting that a police presence inhibits speeding behaviour in both groups.

Conclusion

The research presented demonstrates the use of the TPB within a specific context and suggests that the prediction and understanding of speeding behaviour of young drivers needs to be further investigated. While the TPB might provide some basis for such an investigation, the results of this research suggest that speeding behaviour is not intentional. While young drivers do not intend to speed, they expect to speed. The lack of intention to speed means that interventions to address this behaviour need to be conducted through public media campaigns, changes to the driving environment, and enforcement. It appears that individual interventions are limited to time management, and planning of trips to avoid time pressures to speed rather than motivational issues.

It may be useful for researchers wishing to develop TPB methodology to consider additional relevant variables that will enhance the prediction and understanding of young driver speeding behaviour. The issue of an emotional aspect to this behaviour is an example of such a variable that should be considered in future investigations.

References

Ajzen, I. (1991). The theory of planned behavior. *Organizational Behavior and Human Decision Processes, 50*, 179–211.

Ajzen, I. (2006). Constructing a TPB questionnaire: Conceptual and methodological considerations. Retrieved 27 July 2009, from http://people.umass.edu/aizen/pdf/tpb.measurement.pdf

Ajzen, I., and Fishbein, M. (1980). *Understanding Attitudes and Predicting Social Behavior*. Englewood Cliffs, NJ: Prentice-Hall.

Ajzen, I., and Fishbein, M. (2005). The Influence of Attitudes on Behaviour. In D. Albarracin, B.T. Johnson and M.P. Zanna (Eds.), *Handbook of Attitudes* (pp. 173–222). Mahwah, NJ: Lawrence Erlbaum Associates.

Armitage, C.J., and Conner, M. (2001). Efficacy of the theory of planned behaviour: A meta-analytic review. *British Journal of Social Psychology, 40*(4), 471–499.

Carcary, W.B. (2002). *The effectiveness of pre-driver training*. Paper presented at the 67th Road Safety Congress, Safer Driving – The road to success. 4–6 March, Stratford upon Avon, Warwickshire.

Chatzisarantis, N.L.D., and Hagger, M.S. (2007). Mindfulness and the intention-behavior relationship within the theory of planned behavior. *Personality and Social Psychology Bulletin, 33*(5), 663–676.

Clarke, D.D., Ward, P., Bartle, C., and Truman, W. (2006). Young driver accidents in the UK: The influence of age, experience, and time of day. *Accident Analysis and Prevention, 38*, 871–78.

Cohen, J. (1992). A power primer. *Psychological Bulletin, 112*, 155–59.

Conner, M., Lawton, R., Parker, D., Chorlton, K., Manstead, A.S.R., and Stradling, S. (2007). Application of the theory of planned behaviour to the prediction of objectively assessed breaking of posted speed limits. *British Journal of Psychology, 98*(3), 429–53.

Deery, H.A., and Love, A.W. (1996). The driving expectancy questionnaire: Development, psychometric assessment and predictive utility among young drink-drivers. *Journal of Studies on Alcohol, 57*(2), 193–202.

DETR (2000). New directions in speed management: A review of policy. London: Department of the Environment, Transport, and the Regions.

Elliott, M.A. (2010). Predicting motorcyclists' intentions to speed: Effects of selected cognitions from the theory of planned behaviour, self-identity and social identity. *Accident Analysis and Prevention, 4*, 718–25.

Ferguson, C.A. (2005). *An evaluation of a predriver training program using the Theory of Planned Behaviour (TPB)*. Paper presented at the Australasian Road Safety Research, Policing and Education Conference, Wellington, New Zealand.

Fergusson, D., Swain-Campbell, N., and Horwood, J. (2003). Risky driving behaviour in young people: Prevalence, personal characteristics and traffic accidents. *Australian and New Zealand Journal of Public Health, 27*(3), 337–42.

Gibbons, F.X., Gerrard, M., Blanton, H., and Russell, D.W. (1998). Reasoned action and social reaction: Willingness and intention as independent predictors of health risk. *Journal of Personality and Social Psychology, 74*(5), 1164–80.

Hatfield, J., and Job, R.F.S. (2006). *Beliefs and Attitudes about Speeding*. Canberra: ATSB.

Mitchell-Taverner, P., Zipparo, L., and Goldsworthy, J. (2003). *Survey on Speeding and Enforcement*. Civic Square, ACT: ATSB.

Nagelkerke, N.J.D. (1991). A note on a general definition of the coefficient of determination. *Biometrika, 78*(3), 691–92.

Sheeran, P. (2002). Intention-behaviour relations: A conceptual and empirical review. *European Review of Social Psychology, 12*, 1–36.

Sutton, S. (1998). Predicting and explaining intentions and behaviour: How well are we doing? *Journal of Applied Social Psychology, 28*(15), 1317–38.

Vassallo, S., Smart, D., Cockfield, S., Gunatillake, T., Harris, A., and Harrison, W. (2010). *In the driver's seat II: Beyond the early driving years*. Melbourne: Australian Institute of Family Studies Research Report No 17.

Chapter 15

Older Drivers' Hazard Perception Performance

Tania Dukic
VTI Swedish National Road and Transport Research Institute, Sweden

Emelie Eriksson
Swedish Transport Agency, Sweden

Fridulv Sagberg
Institute of Transport Economics, Norway

Introduction

The population is growing all over the world and humans live longer and healthier lives than just a generation ago (Coughlin, 2007). Also, the population's mobility has increased with globalization. These two aspects implicate a larger number of older drivers, which presents a great challenge to develop and adapt our society to maintain safety (Rosenbloom, 2004). Previous research has found that typically crashes involving older driver often include another vehicle and occur at left-turn intersections with heavy traffic, which impose a high visual load and substantial demands on perceptive and cognitive skills (Oxley et al., 2006; Skying et al., 2009; Viano and Ridella, 1996; Villalba et al., 2001).

Impairment of hazard perception skills has been suggested to be a determinant of crash occurrence. The ability to anticipate traffic situations and other road users' behaviour is an important aspect of safe and efficient driving. Previous hazard perception literature has focused on novice and inexperienced drivers (Wallis and Horswill, 2007), although hazard perception could also be relevant to study for older drivers in order to explain their involvement in some traffic situations.

In a driving situation, the visual channel is the primary source of sensory input. Visual capacity changes among older people have been widely documented both on the physiological level and on the functional level. The question is raised whether visual capacity might explain observed differences in the ability to anticipate hazardous situations in traffic.

The overall objective of the study was to investigate whether visual search behaviours differ between age groups, and if these differences depend on hazard type. The purpose of the study was to provide more knowledge about possible difficulties among older drivers concerning the perception and interpretation of traffic hazards.

Method

A total of 50 participants performed a hazard perception test. They were divided into two groups, one group of middle age drivers (35–55 years old), and one group of older drivers (65 years old and above). The hazard perception test for the present study was a video-based test developed by Sagberg and Bjørnskau (2006). The test consists of 13 traffic scenes recorded from the drivers' viewpoint while driving in various conditions. A critical situation in the test is defined as 'any motion by some of the other road users, which could possibly develop into a hazard, and for which the driver had to be especially prepared for braking or steering'. For each situation a critical interval was defined, during which a response was considered relevant for the situation. Participants were instructed to push the space key on the keyboard as soon as possible whenever a possible hazardous situation was detected. For those who responded to a situation, reaction time was automatically computed for the first response, if several were made during each critical interval.

The first step in the analysis process was to summarize whether participants responded to the hazardous situation or not. Comparisons between age groups were made for reaction time and the number of participants that responded.

Participants who performed the hazard perception test were wearing an eye tracker. The Iview® eye tracker from SMI was used to record eye movements during the hazard perception test. The sampling frequency used was 50 Hz. Visual behaviour was analysed by areas of interest (AOI) and by the amount of time spent on each AOI. Four areas of interest were selected: the hazard object, other road users (both pedestrians and other vehicles), road information (such as signs and traffic lights) and the environment (objects not relevant for driving such as buildings).

Results

Hazard Classification

The hazards in the 13 videos included in the test were classified according to the available literature. Three kinds of hazards were selected and defined.

Firstly, an *obvious hazard* consists of a moving object usually in front of the car, for example a pedestrian walking across the street. If the driver does not take any evasive action such as braking or steering, an accident or a traffic conflict could be expected to happen (Renge et al., 2005) (see Figure 15.1).

Secondly, a *context hazard*, which is also defined as predicting behaviour by Crundall et al. (2008) or a hazard relating to the prediction of other road users' behaviour by Renge et al. (2005), consists of a standing or slowly moving road user on the side of the street. An example of this type of hazard would be pedestrians or slowly moving vehicles on the left or right side of an intersection which can be dangerous, depending on their potential unexpected behaviour. Drivers should anticipate the possible hazards that may develop from such behaviour in order to avoid a potential collision with them (Crundall et al., 2008; Renge et al., 2005; Vlakveld and Twisk, 2008) (see Figure 15.2).

Figure 15.1   An obvious hazard (car ahead has to stop to back into parking space)

Figure 15.2   A context hazard (woman on right pavement walks towards the pedestrian crossing)

The third class of hazard, *context/hidden hazards*, are defined in the literature as hazard prediction from the environment (Crundall et al., 2008) or potential hazards (Renge et al., 2005), which consist of obstacles that obstruct the vision, like a parked vehicle or an intersection. Those hazards can be anticipated to appear from a blind spot in the environment (see Figure 15.3).

Figure 15.3   Context/hidden hazard (pedestrians on the left hidden by an oncoming van)

The results show that the reaction times in general were longer for older drivers than for the younger age group; they were significantly longer for older participants in 5 out of the 13 situations (Table 15.1). However, a larger variation in reaction time was observed for older participants compared with younger ones.

Table 15.1   Mean hazard perception reaction time (in seconds) for the 13 situations, by age group

Situation	Older group	Younger group	p. value
1	2.68	2.03	0.012
2	2.80	2.31	
3	6.41	3.63	< 0.001
4	7.25	7.12	
5	8.79	3.87	< 0.001
6	3.10	3.41	
7	3.83	3.76	
8	2.86	2.42	
9	2.63	2.09	< 0.001
10	2.87	2.24	0.013
11	2.40	2.11	
12	3.60	3.02	
13	1.91	1.86	

Eye Movements

Eye movement analysis was based upon situations where the quality of the data was good enough to analyse, which included 10 out of 13 video recordings. In 6 of the 10 situations the majority of the respondents in both age groups reacted to the hazards (see Figure 15.4). However, in the other four remaining situations less than 50 per cent of the older participants reacted to the hazard, even though they fixated on them, while the majority of the younger participants reacted to the hazards in these four situations. These results indicate that the majority of the older drivers interpreted the predefined hazards as non-hazardous, while the majority of the younger drivers interpreted them as hazardous. The differences were however not significant. Looking closely at the nature of these four situations, it was observed that all situations were comprised of contextual hazards in the form of pedestrians or cyclists.

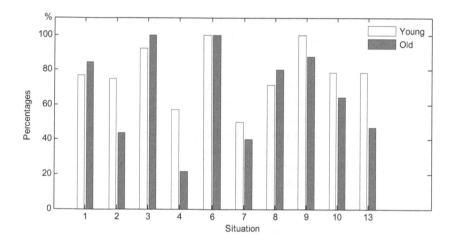

Figure 15.4   Percentages of participants in each age group that reacted to the hazard in each situation

Visual Behaviour

Visual behaviour was analysed further for the situations where the majority of the participants reacted to the hazards by pressing the space key. Hence, in total, six of the situations were further analysed. The total fixation duration distribution on each AOI is summarized in Figure 15.5. Overall, participants spent most of the time looking at the hazard objects as well as other road users. Information concerning the road itself and the environmental objects were not fixated on to the same extent.

Advances in Traffic Psychology

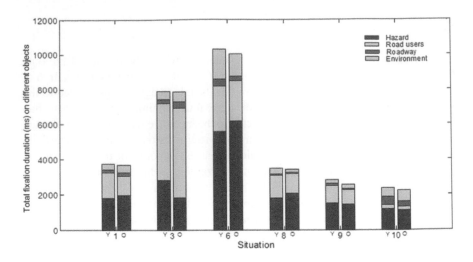

Figure 15.5   Total fixation duration time on each AOI (in ms) by age group
(Y = young; O = old) and situation

The younger age group fixated on the hazard objects to a greater extent than the older age group in situations 3 (context/hidden hazard), 9 (context hazard) and 10 (obvious hazard), while the results were reversed for situations 1 (context hazard), 6 (obvious hazard) and 8 (obvious hazard). The relative differences, when watching the hazard object, were greatest in situation 3.

The between-subject factor age did not show any significant differences when investigating the total fixation duration on the hazard object, while the within-subject factor hazard class did ($F = 236.7$, $p < 0.001$) (Figure 15.6). The interaction effect between these factors showed significant differences between age groups ($F = 3.9$, $p = 0.035$). A closer look at the interaction effect showed that it was the class of context/hidden hazards which differed significantly between age groups ($p = 0.002$). The context/hidden hazards were depicted only in situation 3.

Overall, a significant difference in fixation duration on the hazard object was found for the situation composed of a contextual/hidden hazard. Older drivers had significantly lower total fixation times on the contextual/hidden situation and looked at objects in a narrower field of view. For this situation, mean fixation duration was also longer for younger drivers, although this was not significant. A closer look at situation 3 showed that, compared with the younger group, the older group did not perceive the hazard to a greater extent until the hazard became visible from behind the van (Figure 15.3), even though the hazard was visible earlier.

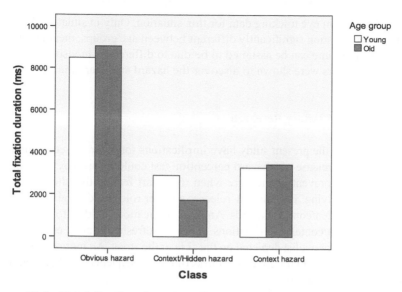

Figure 15.6 **Total fixation duration (in ms) on hazards by age group and class of hazards**

Discussion

For four of the situations there was a tendency among the older drivers not to interpret the situations as hazardous, in other words they did not press the response button even though they fixated on the hazards in these situations. Most of the younger drivers on the other hand reacted to the hazards. In particular, contextual hazards, which were composed of pedestrians or cyclists, were interpreted as significantly less hazardous by older drivers compared with younger drivers. Hence, older drivers did not seem to interpret the contextual hazards as being as risky as younger drivers did.

There were significant differences in total fixation duration time between age groups for the context/hidden hazard for situation 3 only (see Figure 15.3), where older drivers did not identify the hazardous object as early as the younger participants. The hazard was identified much later by the majority of the older drivers in comparison to the younger age group, something that could possibly be explained by the older participants' reduced Useful Field Of View (UFOV). Among the situations analysed in this study, the differences in reaction times between age groups were also shown to be greatest in situation 3 (see reaction times in Table 15.1). Table 15.1 also shows that the older age group reacted significantly slower to the predefined hazards in situations 1, 3, 5, 9 and 10, compared with the younger age group. Probably, the slower reaction time among older drivers was primarily due to cognitive factors in situations 1, 9 and 10, rather than differences in visual behaviour between the age groups. Situation 5 has not been further analysed due to a loss of quality in eye tracking data for this situation. Only in situation 3 was the total fixation duration significantly different between age groups, meaning that the slower reaction time can be assumed to be due to differences in visual behaviour, since older drivers were shown to discover the hazard at a later time.

Conclusions and Future Research

The results from the present study have implications for older drivers' future safe mobility. A new release of a hazard perception test could be used by older drivers themselves as a preventive measure when they start feeling that they are less in control whilst driving. For the new release, a larger role could be given to videos representing hidden/context hazards. Another application could be to expose older drivers to hidden/contextual situations during a refresher course, especially one with pedestrians or cyclist depicted on the side of the street. An increased exposure to such situations could improve awareness of the consequences of their behaviour and prolong safe mobility.

References

Coughlin, J. (2007). Disruptive demographics, design and the future of everyday environments. *Design Management Review*, *18*, 53–9.

Crundall, D., Chapman, P., Trawley, S., and Underwood, G. (2008). *Some hazards are more attractive than others*. Paper presented at the Proceedings from 4th International Conference on Traffic and Transport Psychology, Washington DC, USA, 31 August – 4 September.

Oxley, J., Fildes, B., Corben, B., and Langford, J. (2006). Intersection design for older drivers. *Transportation Research Part F*, *9*(5), 335–46.

Renge, K., Ishibashi, T., Oiri, M., Ota, H., Tsunenari, S., and Mukai, M. (2005). The elderly drivers' hazard perception and driving performance. In G. Underwood (Ed.), *Traffic and Transport Psychology: Theory and Application* (pp. 91–100). Oxford: Elsevier.

Rosenbloom, S. (2004). Mobility of the Elderly. In A.J.S. Clarke. (Ed.), *Transportation in an Aging Society* (pp. 3–21). Washington: Transportation Research Board.

Sagberg, F., and Bjornskau, T. (2006). Hazard perception and driving experience among novice drivers. *Accident Analysis and Prevention*, *38*, 407–14.

Skying, M., Berg, H.-Y., and Laflamme, L. (2009). A pattern analysis of traffic crashes fatal to older drivers. *Accident Analysis and Prevention*, *41*, 253–8.

Wallis, T.S.A., and Horswill, M.S. (2007). Using fuzzy signal detection theory to determine why experienced and trained drivers respond faster than novices in a hazard perception test. *Accident Analysis and Prevention*, *39*, 1137–85.

Viano, D.C., and Ridella, S. (1996). Significance of intersection crashes for older drivers. *Paper 960457*, Warrendale, PA: Society of Automotive Engineers.

Villalba, J., Kirk, A., and Stamadiadis, N. (2001). *Effects of age and cohort on older drivers*: SAE.

Vlakveld, W., and Twisk, D. (2008, 31 August – 4 September). *Young novice drivers, their performance on PC based hazard perception tasks and their crash rate*. Paper presented at the 4th International Conference on Traffic and Transport Psychology, Washington DC, USA.

Chapter 16

Predicting Traffic Accident Rates: Human Values Add Predictive Power to Age and Gender

Ivars Austers, Viesturs Renge and Inese Muzikante

Department of Psychology, University of Latvia

Introduction

It has been well documented by researchers in traffic psychology that age and gender can help in the prediction of accident rates – males and younger drivers tend to be involved in more accidents (for example, Krahé, 2005; Krahé and Fenske, 2002; Stradling and Parker, 1997). There are a number of studies showing that driving style and a tendency towards specific deviations from safe driving patterns is inherited from parents – children display the same driving style as their parents (Bianchi and Summala, 2004; Taubman-Ben-Ari et al., 2005). This may provide support for arguments that there are other culturally shared motivational forces in addition to the family that shape driving behaviour. Our proposal in the present paper is that human values carry significant power in explanations of deviations from safe driving patterns. The role of values in this context has been studied on a cultural (in other words country) level (Özkan and Lajunen, 2007). There have also been studies which relate values to driving behaviour in general, for example values have been used in explaining socially responsible car usage. It has been shown that egocentric values are positively related to socially problematic usage of transport, while altruistic and biospheric values related negatively (de Groot and Steg, 2007). To our knowledge, there have been two studies of a smaller scale thus far, where individual values were treated as predictor variables in explaining traffic accidents (Muzikante and Renge, 2008; Renge et al., 2008). However, these two studies were primarily focused on other research issues besides values and involved relatively small samples of drivers. In the present study our aim was to test whether human values, in addition to age and gender, carry incremental validity in explaining traffic accidents.

Recent studies in psychology generally use value theory as developed by Shalom Schwartz (Bardi and Schwartz, 2003; Schwartz, 1992), where human values are defined as a set of abstract beliefs related to desired human goals. The survey in 67 countries showed that values carry a universal structure, characteristic of all people, but there may be differences in the hierarchy of these values

(Schwartz et al., 2001). People may differ markedly depending on the significance they ascribe to each of the 10 basic value types, yet the basic structure of the value types and motivational concerns they represent (in other words the conflict and compatibility of the values) remain the same independent of the culture that the people represent (Schwartz, 1992).

All the 10 value types form two concurrent dimensions (Figure 16.1). The values from the opposite sides of the circle are in conflict, while the adjacent values are similar in their content and motivational goals. The first dimension, 'Openness to Change/Conservation', consists of stimulation, hedonism and self-direction versus security, traditions and conformity in values. The first two values stress independence in actions, thoughts and feelings and openness to new experiences, whereas the last two stress self-restriction, orderliness and resistance to change. The second dimension is 'Self-Enhancement/Self-Transcendence' – power, hedonism and achievement versus universalism and benevolence. 'Openness to Change' and 'Self-Enhancement' emphasizes the realization of one's own interests, while 'Self-Transcendence' and 'Conservation' care about the interests and welfare of others. If one acts to fulfil any value, it may lead to psychological, practical or social consequences which may not always be in accord with the fulfilment of another value; in other words, they may be in conflict (Schwartz, 1992). Conflict between two concurrent values is generally solved in favour of the value which is most significant at the given moment.

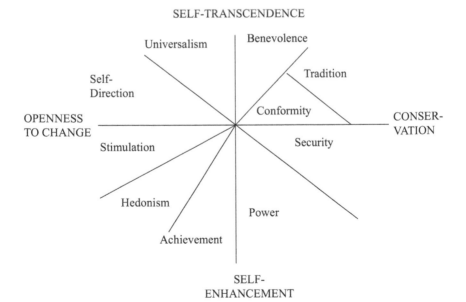

Figure 16.1   Circular value structure (Schwartz, 1992)

A significant question in the values area is whether values influence behaviour. Doubts about the existence of such an influence have been among the reasons why historically relatively little attention has been paid to the study of values. There have been studies showing that there is only a moderate or sometimes even a weak relationship between individual values and different types of social behaviour (Feather and O'Brien, 1987; Henry, 1976; Leung et al., 1995). However, more recent studies (Verplanken and Holland, 2002; Lönnqvist et al., 2006) have tried to determine the conditions which allow values to predict behaviour. According to Verplanken and Holland (2002), only the so-called central values will be able to influence behaviour. Values will achieve a central status when they are a part of an individual's self- identity. For instance, in a case of 'value centrality', a person identifying him or herself as an environmental activist will be using values that form part of his or her identity. Verplanken and Holland's research demonstrated that values will predict behaviour if they are cognitively activated and central to a person. Cognitive activation of values takes place when these values become salient under the focus of attention, for example, in a discussion of the violation of traffic rules in the light of the importance of driving safety. Values also may be activated by a confrontational situation. For example, a driver of a sports car has to make a choice – whether to demonstrate the maximum speed of a car to his or her friends or to drive in a way that ensures the safety of the friends and all others involved in traffic; stimulation and security values will be in conflict here (Verplanken and Holland, 2002). Research has also demonstrated that if high importance is attached to conformity values, this will lead an individual to behave according to social norms, but those for whom conformity is less important have a tendency to care more about their individual values (Lönnqvist et al., 2006).

According to the classification proposed by the Manchester group of researchers (Reason et al., 1990; Parker et al., 1995) all the possible deviations in driving will fall into one of three types: *Violations*, *Errors* and *Lapses*. *Violations* are intended deviations from the rules of safe driving (for example 'Deliberately disregard the speed limits late at night'), *Errors* are unsuccessful efforts to reach a chosen goal (for example 'Misjudge the speed of an oncoming vehicle'), but *Lapses* are minor deviations on the way to the chosen goal. Since *Violations* are intentional actions, they are related to social and motivational factors, whereas *Errors* and *Lapses* depend on a driver's informational processing. For the assessment of these factors, a Driver Behaviour Questionnaire (DBQ) was developed. Local adaptations of the DBQ are being used in different countries (for example, Åberg and Rimmö, 1998; Obriot-Claudel and Gabaude, 2004). Several studies have demonstrated that males are more prone to *Violations*, but females – to *Lapses*. In addition, *Violations* decrease with driver age, but not *Errors*. Of the three types of deviant driving, conscious *Violations* have the closest relationship to traffic accidents (for example, Lajunen et al., 2004; Mesken et al., 2002; Parker et al., 1995; Reason et al., 1990; Underwood et al., 1997).

We made the following predictions concerning the relationship of aberrant driving and accident rate based on the importance of individual values. First, based on previous studies (Muzikante and Renge, 2008; Renge et al., 2008), we expected that individual values would not predict the number of accidents directly, but that values would predict the extent to which drivers engage in *Violations*. Values were expected to influence the accident rate indirectly via *Violations*. Besides gender (males) and age (younger drivers) being more prone to *Violations*, we also expected that values would add a predictive power to age and gender. Specifically – 'Self-Enhancement' values were expected to positively relate to *Violations*, while 'Conservation' and 'Self-Transcendence' values were expected to relate negatively to *Violations*. We did not formulate a specific prediction with respect to 'Openness to Change' values.

Method

Participants

There were 530 participants in the study. The mean age of the respondents was 29.71 (SD = 8.36, the range 18 to 60, 46 per cent were females). Forty per cent of the participants had been involved in accidents in the previous three years.

Measures

> *Individual values*: For the assessment of individual values, the Latvian version of the Portrait Values Questionnaire (PVQ; Schwartz et al., 2001) was used. We computed four index variables corresponding to the four dimensions of the values system, namely, 'Self-Transcendence', 'Self-Enhancement', 'Openness to Change' and 'Conservation'. The reliability of all the four indexes was satisfactory (Cronbach's alphas ranged from .54 to .68).
>
> *Risky driving*: The Latvian version of the Driver Behaviour Questionnaire was used in this study. It was created using items from the original (Reason et al., 1990) and Swedish versions of the scale (Åberg and Rimmö, 1998). The questionnaire consisted of 29 items related to different driving situations) and was answered on a six-point Likert scale (1 – never, 6 – nearly always). The reliability (measured by Cronbach's alpha) of the subscales were reasonable high: .79 for *Violations* .69 for *Errors*, and .70 for *Lapses*.

The respondents were also asked to indicate the frequency of involvement in road traffic accidents (during the last three years) and their age and gender.

Procedure

An Internet-based version of the questionnaire was used (posted at a portal dedicated to automobiles and driving), which was filled out by 545 respondents. Data analysis was performed only on a set of 530 questionnaires, since 15 questionnaires had too much missing data.

Results

Correlations among the major variables (including age and gender) are represented in Table 16.1. Age significantly correlated with *Violations, Errors* and accident involvement. Gender significantly correlated with *Violations, Lapses* and the number of accidents in the last three years. Males were more prone to report *Violations* and accidents, while women were more prone to report *Lapses*. Age was negatively related to accidents.

A hierarchical regression analysis was conducted to determine the incremental validity of values over and above age and gender in predicting *Violations*. Table 16.2 shows the results of the regression analysis. As can be seen, values were significant predictors of *Violations* over and above age and gender. Adding values in Step 2 increased the explained variation in *Violations* by 9 per cent over that explained by age and gender, while together they explained 22 per cent of the variance in *Violations*. As we hypothesized, 'Self-Enhancement' values positively predicted *Violations* and 'Conservation' values negatively predicted *Violations*. 'Self-Transcendence' values along with 'Openness to Change' values did not predict *Violations*.

Table 16.1   Means, standard deviations and Pearson correlations among variables

	M	SD	Sex	Age	Vio	Errors	Lapses	Accident
Violations	2.76	0.78	.24**	- .28**	-	-	-	-
Errors	1.72	0.47	.03	- .20**	.43**	-	-	-
Lapses	2.01	0.54	- .22**	- .01	.24**	.50**	-	-
Accident involvement	0.39	0.49	.18**	- .09*	.14**	.08	.03	-
Self enhancement	3.68	1.05	- .01	-.25**	.25**	.15**	.14**	.04
Openness to change	3.99	0.99	.03	- .12*	.17**	.06	.03	.12*
Self- Transcendence	4.14	0.91	- .16**	.17**	- .20**	- .17**	- .01	.03
Conservation	3.86	0.97	- .13**	.21**	- .31**	- .20**	- .01	-.04

* p < .05, ** p < .01

Advances in Traffic Psychology

To test the model of the relationship of individual values to *Violations* and drivers' accident involvement, we conducted Path Analysis using a series of regression analyses. Table 16.2 presents the first regression analysis and Figure 16.2 illustrates all the significant paths (only statistically significant standardized regression coefficients are presented) identified by the analysis.

Those findings were in line with the predictions. On the other hand, 'Openness to Change' and 'Self Transcendence' values had no influence on *Violations*. In addition, there were no direct influences of values on involvement in accidents. However, *Violations* predicted the number of accidents (standardized regression coefficient = .16).

Table 16.2   Summary of hierarchical regression analysis of age, gender and values on violations

Independent variables	B	SE B	β	R^2	ΔR^2	F change
Step 1				.13	.13	38.78***
Age	- 0.02	0.01	- .27***			
Gender	0.35	0.06	.23***			
Step 2			- .16***	.22	.09	15.23***
			.20***			
Age	- 0.01	0.01	.18***			
Gender	0.30	0.06				
Self - Enhancement	0.13	0.03	.07			
Openness to Change						
Self - Transcendence	0.06	0.03	- .05			
Conservation						
	- 0.04	0.04	- .22***			
	- 0.17	0.04				

*** $p < .001$

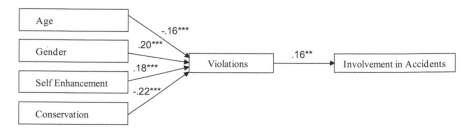

** $p < .01$, *** $p < .001$

Figure 16.2   Computed model linking human values to number of accidents via violations

Discussion

One of the first studies at the very beginnings of traffic psychology which related personality factors to the risk of traffic accidents – the famous research about taxi drivers – found that mild social deviance was related to involvement in traffic accidents (Tillman and Hobbs, 1949, cited in Lawton et al., 1997). As a result of this study the well-known expression 'We drive as we live' was coined. Later studies have confirmed this relationship (Lawton et al., 1997; Underwood et al., 1997; West et al., 1993).

Values influence our lives in many ways, as one may conclude from the results of the present study; they also influence our driving style. The predictions of our study were supported to a large extent. We found that 'Self-Enhancement' values positively predicted *Violations* and 'Conservation' values negatively predicted traffic *Violations*. However, the model did not show a direct influence of those variables on the accident rate. The only direct predictor of accidents was traffic *Violations*, suggesting that *Violations* mediate the relationship between the culturally embedded variables and accident involvement.

The finding that 'Self-Enhancement' values correlate with driving *Violations* is consistent with the results of research done by Bond and his colleagues (Bond et al., 2004). They found that 'Self-Enhancement' values correlate with competition and wishful thinking. Wishful thinking is related to risky driving and traffic accidents (Kontogiannis, 2006) and competition as a strategy is a fairly aggressive way of conflict resolution. Aggressiveness in turn is related to traffic accidents (for example, Lajunen and Parker, 2001; Van Rooy et al., 2006). The negative relationship between 'Conservation' values and *Violations* corresponded to the results in a cross-cultural study at a country level where a negative correlation between 'Conservation' and traffic safety was found (Özkan and Lajunen, 2007).

However, our results are inconsistent with a previous study's results to some extent. In Bond et al's. (2004) study, 'Openness to Change' values were negatively related to wishful thinking, but in our study they had a positive correlation with driving *Violations* and traffic accident involvement. However, they did not appear as a significant independent variable in the final path analysis model.

All in all, the findings of the present study demonstrate that the idea of combining age, gender and values is also fruitful in studying aberrant driving behaviour. So, from this we can conclude that individual values do not predict the frequency of involvement in traffic accidents per se. Value dimensions of 'Conservation' and 'Self Enhancement' are significant predictors of the traffic accident rate in addition to age and gender; whereas, the frequency of *Violations* predicts the involvement in traffic accidents. Our conclusion is that *Violations* act as a mediating variable between personality related variables (individual values, age and gender) and involvement in traffic accidents.

However, the data in the present study do not provide a full understanding of the mechanisms underlying the demonstrated relationship and the role of *Violations* as a mediating variable. It is possible that studies so far have not considered the idea and the importance of self-centrality (Verplanken and Holland, 2002).

Acknowledgments

This research was supported by Grant No. 09.1508 from the Latvian Council of Science.

References

Åberg, L., and Rimmö, P.-A. (1998). Dimensions of Aberrant Driver Behaviour. *Ergonomics, 41*, 39–56.

Bardi, A., and Schwartz, S.H. (2003). Values and Behaviour: Strength and Structure of Relations. *Personality and Social Psychology Bulletin, 29*, 1207–20.

Bianchi, A., and Summala, H. (2004). The 'Genetics' of Driving Behaviour: Parent Driving Style Predicts Their Children's Driving Style. *Accident Analysis and Prevention, 36*, 655–9.

Bond, M.H., Leung, K., Au, A., Tong, K-K., and Chemonges-Nielson, Z. (2004). Combining Social Axioms with Values in Predicting Social Behaviours. *European Journal of Personality, 18*, 177–91.

Feather, N.T., and O'Brien, G.E. (1987). Looking for Employment: An Expectancy-Valence Analysis of Job-Seeking Behaviour among Young People. *British Journal of Psychology, 78*, 251–71.

de Groot, J.I.M., and Steg, L. (2007). Values, Beliefs and Environmental Behaviour: Validation of an Instrument to Measure Egoistic, Altruistic and Biospheric Value Orientations in Five Countries. *Journal of Cross Cultural Psychology, 38*, 318–32.

Henry, W.A. (1976). Cultural Values do Correlate with Consumer Behavior. *Journal of Marketing Research, 13*, 121–7.

Krahé, B. (2005). Predictors of Womens' Aggressive Driving Behavior. *Aggressive Behavior, 31*, 537–46.

Krahé, B., and Fenske, I. (2002). Predicting Aggressive Driving Behavior: The Role of Macho Personality, Age, and Power of Car. *Aggressive Behavior, 28*, 21–9.

Kontogiannis, T. (2006). Patterns of Driver Stress and Coping Strategies in a Greek Sample and Their Relationship to Aberrant Behaviors and Traffic Accidents. *Accident Analysis and Prevention, 38*, 913–24.

Lajunen, T., and Parker, D. (2001). Are Aggressive People Aggressive Drivers? A Study of the Relationship between Self-Reported General Aggressiveness, Driver Anger and Aggressive Driving. *Accident Analysis and Prevention, 33*, 243–55.

Lajunen, T., Parker, D., and Summala, H. (2004). The Manchester Driver Behaviour Questionnaire: A Cross-Cultural Study. *Accident Analysis and Prevention, 36*, 231–8.

Lawton, R., Parker, D., Stradling, S.G., and Manstead, A.S.R. (1997). Predicting Road Traffic Accidents: The Role of Social Deviance and Violations. *British Journal of Psychology, 88*, 249–62.

Leung, K., Bond, M.H., and Schwartz, S.H. (1995). How to Explain Cross-Cultural Differences: Values, Valences, and Expectancies? *Asian Journal of Psychology, 1*, 70–5.

Lönnqvist, J.-E., Leikas, S., Paunonen, S., Nissinenm V., and Verkasala, M. (2006). Conformism Moderates the Relations between Values, Anticipated Regret, and Behavior. *Personality and Social Psychology Bulletin, 32*, 1469–81.

Mesken, J., Lajunen, T., and Summala, H. (2002). Interpersonal Violations, Speeding Violations and Their Relation to Accident Involvement in Finland. *Ergonomics, 45*, 469–83.

Muzikante, I., and Renge, V. (2008). Autovaditāju Individualo Vertibu Saistiba ar Riskantu Brauksanu [Drivers' Individual Values and Risky Driving], *LU.* Rakstu *Krajums. Psihologija*, 742: 39–55, Riga: Akademiskais apgads.

Obriot-Claudel, F., and Gabaude, C. (2004). The Driver Behaviour Questionnaire: A French Study Applied to Elderly Drivers. Paper presented at the *3rd International Conference on Traffic and Transport Psychology*, Nottingham, UK. Retrieved from www.psychology.nottingham.ac.uk/IAAPdiv13/ICTTP2004papers2/Individual per cent20Differences/Obriot.pdf

Özkan, T., and Lajunen, T. (2007). The Role of Personality, Culture, and Economy in Unintentional Fatalities: An Aggregated Level Analysis. *Personality and Individual Differences, 43*, 519–30.

Parker, D., West, R., Stradling, S., and Manstead, A.S. (1995). Behavioural Characteristics and Involvement in Different Types of Traffic Accident. *Accident Analysis and Prevention, 27*, 571–81.

Reason, J.T., Manstead, A., Stradling, S., Baxter, J., and Campbell, K. (1990). Errors and Violations on the Road: A Real Distinction? *Ergonomics, 33*, 1315–32.

Renge, V., Austers, I., and Muzikante, I. (2008). Combining Social Axioms with Personality Measures and Self-Reported Driving Behaviour in Predicting Traffic Accidents. *International Journal of Psychology, 43*, 586.

Schwartz, S.H. (1992). Universals in the Content and Structure of Values: Theory and Empirical Tests in 20 Countries. In M. Zanna (Ed.), *Advances in Experimental Social Psychology, 25*: 1–65. New York: Academic Press.

Schwartz, S.H., Melech, G., Lehmann, A., Burgess, S., and Harris, M. (2001). Extending the Cross-Cultural Validity of the Theory of Basic Human Values with a Different Method of Measurement. *Journal of Cross-Cultural Psychology, 32*, 519–42.

Stradling, S.G., and Parker, D. (1997). Extending the Theory of Planned Behaviour: The Role of Personal Norm, Instrumental Beliefs and Affective Beliefs in Predicting Driving Violation. In T. Rothengatter and E. Carbonell

Vaya (Eds.), *Traffic and Transport Psychology: Theory and Application.* Amsterdam: Pergamon.

Taubman-Ben-Ari, O., Mikulincer, M., and Gillath, O. (2005). From Parents to Children: Similarity in Parents and Offspring Driving Styles. *Transportation Research Part F: Traffic Psychology and Behaviour, 8,* 19–29.

Underwood, G., Chapman, P. Wright, S., and Crundall, D. (1997). Estimating Accident Liability. In T. Rothengatter and E. Carbonell Vaya (Eds.), *Traffic and Transport Psychology: Theory and Application.* Amsterdam: Pergamon.

Van Rooy, D.L., Rotton, J., and Burns, T.M. (2006). Convergent, Discriminant, and Predictive Validity of Aggressive Driving Inventories: They Drive as They Live. *Aggressive Behavior, 32,* 89–98.

Verplanken, B., and Holland, R.W. (2002). Motivated Decision-Making: Effects of Activation and Self-Centrality of Values on Choices and Behavior. *Journal of Personality and Social Psychology, 82,* 434–47.

West, R., Elander, J., and French, D. (1993). Mild Social Deviance, Type-A Behaviour Pattern and Decision-Making Style as Predictors of Self-Reported Driving Style and Traffic Accident Risk. *British Journal of Psychology, 84,* 207–19.

Chapter 17

Examining the Evidence that Drugs Impair Driving: Some Recent Findings from the Drugs and Driving Research Unit (DDRU) at Swinburne University

Con Stough, Rebecca King and Luke Downey

Drugs and Driving Research Unit (DDRU), Swinburne Centre for Human Psychopharmacology, Swinburne University, Melbourne, Australia

Edward Ogden

Drugs and Driving Research Unit (DDRU), Swinburne Centre for Human Psychopharmacology, Swinburne University, Melbourne, Australia and Victoria Police, Melbourne Australia

Introduction

This chapter reviews findings from studies examining the effects of various drugs on driving. These studies include drugs such as cannabis, amphetamines and ketamine. Of relevance are some recent studies conducted by the Drugs and Driving Research Unit which in 2010 celebrated 10 years of research on this specific topic. Two recent DDRU studies are discussed: (1) a randomized placebo controlled trial in which methamphetamine, MDMA and placebo were administered and driving skills assessed as well as (2) the preliminary results of a study examining culpability of various drugs in seriously injured drivers in Victoria.

Deaths and Serious Injuries on Australian Roads

The total number of people killed on Australian roads in 2006 was 1,601, with a further 31,204 people suffering serious injuries (Australian Transport Safety Bureau, 2007; Berry and Harrison, 2008). Many factors are believed to contribute to traffic accidents including drug and alcohol use, fatigue and speed. Alcohol has long been recognized as a major contributor to fatal road accidents. In 2008 28 per cent of all drivers killed on Victorian roads had a Blood Alcohol Concentration (BAC) of .05 per cent or higher (Traffic Accident Commission, 2009). Alcohol significantly increases a driver's risk of a serious road crash (Longo et al., 2000), with drivers over .05 BAC being more than twice as likely to be responsible for the accident than a driver who is drug and alcohol free (Drummer et al., 2004). The effect of even low doses of alcohol on driving and on the risk of motor accidents has been extensively studied and will not be reviewed in this chapter. Instead we discuss the rather less well-established evidence for the impairment in driving due to a range of licit and illicit drugs.

The Use of Illicit Drugs in Australia

The most-used illicit drug in Australia is cannabis, followed by ecstasy and methamphetamine, with over one in ten (13.4 per cent) Australians aged 14 years and older reporting illicit drug use within the past 12 months (Australian Institute of Health and Welfare, 2008). Although the use of cannabis has declined since 1998, it has been used recently by close to 1 in 10 Australians (9.1 per cent) and is still a serious road safety concern. The use of synthetic drugs such as ecstasy, amphetamine and methamphetamine is steadily increasing around Australia with over one in ten (12.0 per cent) 20–29-year-olds reporting use within the past 12 months. These types of synthetic drugs are gaining popularity and are used predominantly by young people and truck drivers (Drugs and Crime Prevention Committee, 2004).

Due to the high proportion of Australians using illicit drugs there is growing concern over their involvement in traffic accidents and fatalities. Road statistics indicate that drug related road accidents and deaths have increased dramatically over the last 15 years. From 1990 to 1993, 22 per cent of all Victorian drivers killed tested positive for the presence of drugs other than alcohol (9.6 per cent involved cannabis, 3.9 per cent amphetamines and other stimulants, 4.5 per cent benzodiazepines, and 3.3 per cent opioids). This percentage has increased to 33 per cent in 2004 (12.6 per cent involved cannabis, 4.9 per cent amphetamines and other stimulants, 4.4 per cent benzodiazepines, and 7.1 per cent opioids) (Drummer, 1994; Drummer, 1998; Drummer et al., 2003; Traffic Accident Commission, 2009). These statistics are similar to those reported in other countries. For instance, in Washington, data has revealed that the incidence of impairing drugs found in fatally injured drivers has increased significantly over the last decade, from 25 per cent in 1991 to 35 per cent in 2001 (Schwilke et al., 2006). In particular, fatally injured drivers who tested positive for methamphetamine increased a staggering 200 per cent. Schwilke et al. suggests that this is reflective of the general growth in popularity of this class of illicit substance. The number of fatally injured drivers in the U.K who test positive to illicit drugs has also risen from 3 per cent in the mid 1980s to 18 per cent in 1999 (Jackson et al., 2000). France has also seen a significant increase in the presence of impairing drugs in fatally injured drivers between 2001 and 2004, with cannabis increasing from 16.9 per cent to 28.9 per cent, cocaine from .2 per cent to 3.0 per cent and amphetamines from 1.4 per cent to 3.1 per cent (Mura et al., 2006). These statistics demonstrate the significant impact that drug driving has on road safety.

Analysis of drivers injured in road traffic accidents also reveals the significant impact of drugs on road safety. A study which screened blood samples taken from 436 injured drivers attending the Alfred Emergency and Trauma Centre in Prahran, Victoria reported that cannabis metabolites were found in 46.7 per cent of injured drivers, with 7.6 per cent testing positive for delt-9-tetrahydrocannabinol, the active form of cannabis. The next highest drug group detected was benzodiazepines in 15.6 per cent of cases, followed by opiates in 11 per cent, amphetamines in 4.1 per cent, methadone in 3.0 per cent and cocaine in 1.4 per cent. Multiple drug use was also reported in 9.4 per cent of cases with the most common combination between benzodiazepines and opiates, followed by benzodiazepines and cannabis (Ch'ng et al., 2007).

Vic Roads estimate that around 20 per cent of fatal road accidents involve driver fatigue. Many studies have shown that moderate sleep deprivation of 17 to 19 hours produces driving impairments which are equivalent or worse than those seen at .05 per cent BAC (Arnedt et al., 2000; Dawson and Reid, 1997; Falleti et al., 2003; Williamson and Feyer, 2000), while impairments observed at 20 to 24 hours of sleep deprivation have been shown to be equivalent to 0.1 per cent BAC (Dawson and Reid, 1997; Williamson and Feyer, 2000). Fatigue alone is a serious road safety concern, however, when fatigue is combined with substance use such as alcohol or MDMA, driving impairments are even greater than either alone (Arnedt et al., 2000; Kuypers et al., 2006).

Culpability Studies

Some studies have moved beyond stating the mere prevalence of drugs in samples taken from fatally or seriously injured drivers by using methods to assign culpability or responsibility for each accident. This allows us to better understand whether one type of drug is more impairing than another.

Fatal Accidents

In the first of a series of studies, Drummer (1994) collected data for 1,045 drivers killed. Culpability (or responsibility) was determined according to the mitigating factors (independent of drug analysis), and drivers were classified as culpable, contributory, or not culpable. The mitigating factors used in the analyses were the condition of the road and vehicle, driving conditions, type of accident, witness observations, road law obedience, difficulty of the task involved, the level of fatigue (Robertson and Drummer, 1994). The proportion of culpable drivers (odds ratio OR) was calculated for each drug type condition. The large majority (73 per cent) of drivers in the sample as a whole were culpable, while only 18 per cent were not culpable. The highest culpability ratio for any drug alone was found for alcohol. Culpability ratios were considerably higher for all drugs in combination with alcohol, except opiates.

More recently, a large study across Victoria, New South Wales and South Australia revealed that drivers who tested positive to psychoactive drugs were more likely to be responsible for the fatal accident (Drummer et al., 2004). The highest culpability ratio for a single drug was again alcohol. Less than .05 per cent BAC drivers were 1.2 times more likely to be responsible for the accident, while drivers with a BAC over .20 per cent were found to be 24 times more likely to be culpable. Drivers intoxicated with cannabis were also found to be at an increased risk of causing the fatal accident, with THC concentrations over 5ng/ml resulting in a culpability ratio of 6.6 (this is a similar ratio to drivers with a BAC of over .15 per cent). Drivers also frequently combined alcohol and cannabis further increasing the risk of a fatal accident. Stimulants significantly increased a driver's risk of a serious road crash, particularly in truck drivers (culpability ratio of 8.8), while benzodiazepines and opiates were only found to have weak positive relationships with culpability.

Non-Fatal Accidents

Culpability analysis has also been used to determine crash risk for injured drivers. Drivers testing positive to alcohol only, benzodiazepines only, and the combination of alcohol and cannabis and alcohol and benzodiazepines, were significantly more likely to be regarded as culpable for the accident in comparison with those found to be drug and alcohol free. However, unlike results reported by Drummer et al. (2004), drivers who tested positive to cannabis in this study were not found to have an increased crash risk. One explanation for the difference in findings is that lower concentration levels of THC were reported in Longo's study compared to Drummers.

Preliminary data from an ongoing Melbourne study by the DDRU and Victoria Police (Ogden et al., 2009) examining culpability and injured drivers has revealed that alcohol again had the highest culpability ratio for a single drug; at less than .05 per cent BAC drivers were 4.7 times more likely to be responsible for the accident than drug and alcohol free drivers. This increased to an odds ratio of more than 6 for those drivers who were only a little bit over (.05 to .08 per cent BAC), while drivers with a BAC of over .20 per cent were found to be 20 times more likely to be culpable. The most common illicit drug detected was cannabis with 18 per cent of drivers testing positive. Drivers detected with any level of THC had only a slightly larger crash risk than drug free drivers (OR 1.5); however, when THC concentrations were 5ng/ml or more, the risk increased dramatically with an odds ratio of over 6 (Figure 17.1).

Amphetamine type stimulants were detected in 13 per cent of drivers and 95 per cent of these drivers were determined to be responsible for the accident (Figure 17.2).

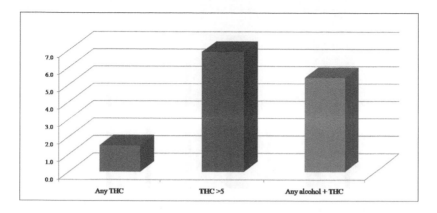

**Figure 17.1   Drugs and culpability analysis from the first 400 accidents in a
new study by Ogden et al. (2009): cannabis**

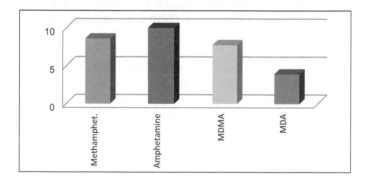

**Figure 17.2   Drugs and culpability analysis from the first 400 accidents in a
new study by Ogden et al. (2009): amphetamines**

Of interest in this study are the high levels of benzodiazepines detected, with
21 per cent of drivers testing positive and 3 per cent with levels in the toxic range.
Most types of benzodiazepines were associated with moderate increases in crash
risk and were found to be dose dependant. Alprazolam in particular stands out with
100 per cent of drivers detected with this benzodiazepine being responsible for the
accident in which they were injured and half of the drivers had toxic levels, clear
evidence of abuse. The data suggests that at therapeutic levels benzodiazepines
have only a slight increase of crash risk, whereas abuse or misuse is associated
with much larger crash risks (Figure 17.3).

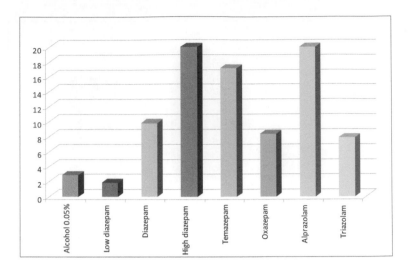

**Figure 17.3   Drugs and culpability analysis from the first 400 accidents in a
new study by Ogden et al. (2009): benzodiazepines**

Poly drug use was also frequently detected with over 20 per cent of drivers
injured in traffic accidents in this sample having more than one drug in their
system. The combinations most often detected were alcohol and cannabis (OR
5.4), alcohol and benzodiazepines (OR 13.8) and cannabis and benzodiazepines
(OR 6.9).

While culpability analyses provide important information concerning the
role of drugs in crashes, they are not conclusive evidence of a causal relationship
because they only include crash-involved drivers. So, while the analyses indicate
that drugs are associated with higher culpability, experimental research is also
needed to examine the link between the presence of these drugs in the blood and
subsequent impairment. In the next section we describe research that has examined
the effect of various drugs on driving and include some direct experimental studies
in which various drugs have been administered to drivers or participants tested on
driving simulators.

Drugs and Driving

Cannabis

Within a few minutes of inhaling cannabis smoke, users often report feeling, along
with intoxication, rapid heartbeat, loss of coordination, poor sense of balance,
slower reaction time, distorted perception, poor concentration and forgetfulness.
Following this acute phase, is the residual or rebound phase where intoxication

subsides rapidly and is characterized by fatigue, excessive sleepiness and inattention.

Research examining driving performance and cannabis indicates a dose dependant relationship, with impairments demonstrated for every performance area related to safe driving: reaction time, tracking, psychomotor skills, visual functions and attention (Berghaus et al., 1995). Simulated and on road driving studies report impaired perceptual processes, such as monitoring the speedometer and maintaining speed; response to stimuli, such as stopping and starting; and subsidiary tasks (Kelly et al., 2004; Ramaekers et al., 2004; Smiley, 1998). Tracking ability is the most consistently reported driving skill to be impaired after cannabis consumption (Ramaekers et al., 2000; Sexton et al., 2000; Smiley, 1986) and is highlighted by an increase in the number of cones hit on a driving course, an increase in sideway movements of the vehicle and an increase in the percentage of time spent out of a lane. In line with these findings, a study funded by the National Drug Law Enforcement Research Fund (NDLERF) at the DDRU at Swinburne University found that the consumption of low dose and high dose cannabis was associated with an increase in vehicle lane weaving (straddling solid and barrier lines) (Papafotiou et al., 2005). Cannabis impairment is most prominent in the first two hours after smoking, however, driving performance has been shown to be compromised up to five hours after cannabis consumption (Berghaus et al., 1995). Drivers can be impaired firstly by cannabis intoxication and later in the residual phase due to fatigue and inattention. Despite common stereotypes regarding the low risk of having an accident under the influence of cannabis, there is no doubt that cannabis poses a significant crash risk for drivers.

Amphetamines

Amphetamines can be taken orally, snorted or injected with users experiencing increased energy levels and mental alertness. The effects intensify with increased doses and users may become over excited, talkative and experience a false sense of self-confidence or superiority. Physical effects of amphetamine can include increased/distorted sensations, hyperactivity, dry mouth, tachycardia, blurred vision, impaired speech, dizziness, uncontrollable movements or shaking. In addition to the physical effects, users often report feeling restless, anxious and moody.

There has been little research to examine the effects of amphetamines on driving performance. Laboratory studies show that at low doses amphetamines improve some skills related to driving (Kelly et al., 2004; Silber et al., 2006). However, due to ethical constraints it is difficult to assess larger recreational doses in the laboratory. Crash statistics indicate that larger doses can lead to driving impairments resulting in death and injury, particularly for truck drivers (Drummer et al., 2004; Ogden et al., 2009). Logan (1996) evaluated actual driving behaviour of drivers arrested or killed in a traffic accident who tested positive for methamphetamine. He reported that drivers often drifted out of the lane of travel, drove erratically, had a tendency to weave, speed, drifted off the road, and were involved in high speed collisions suggesting increased risk taking. The only simulated driving study to date (conducted at the DDRU in Melbourne) showed that dexamphetamine had a deleterious effect on driving performance. Drivers under the dexamphetamine condition were more likely to fail to stop at a red traffic light and signal incorrectly (Silber et al., 2005). A more recent study funded by the Australian Research Council (ARC) at the DDRU (Stough et al., 2007) showed methamphetamine impaired some aspects of driving performance despite improvements in simple reaction time. Similar to cannabis, amphetamine users can experience rebound fatigue due to prolonged wakefulness. More research is needed to evaluate the effects of methamphetamine on driving; in particular it would be useful to determine the extent of rebound fatigue at various doses that approximate community use. However, the rebound effect is likely to cause driving impairment; therefore individuals should not drive at any stage after consuming amphetamines and particularly not drive during the rebound phase. As some amphetamine users also show fatigue (due to staying awake and attending rave and dance parties), the rebound effect is further compounded by fatigue. This latter issue is also observed in MDMA users.

MDMA

An MDMA pill takes effect after 30 to 45 minutes, starting with little rushes of exhilaration, intoxication lasts up to five hours. Along with the positive feelings that the drug is taken for, users may experience disorientation, dry mouth, blurred vision, and involuntary muscular activity.

More research has examined the effects of MDMA on driving performance. MDMA has been demonstrated to improve some aspects of driving such as speed of manual movement (Lamers et al., 2003), tracking (Ramaekers and Kuypers, 2006) and sustained attention (Kuypers et al., 2006) and impair other driving skills such as the ability to perceive and predict motion (Lamers et al., 2003), visual scanning (Parrott and Lasky, 1998) and accuracy of speed adaptation (Ramaekers et al., 2006). A study conducted by De Waard et al. (2000) reported that drivers in the MDMA condition were prepared to accept higher levels of risk. These types of impairment could have serious consequences as drivers manoeuvre in traffic. In contrast to Ramaekers and Kuypers (2006) finding that tracking performance was improved after MDMA, Kuypers et al. (2007) found impairments in tracking ability. The difference between these studies is that Kuypers et al. (2007) administered night-time doses of MDMA in order to more realistically demonstrate what an ecstasy user would experience after taking the drug during the evening and going to a party. The impairments reported were additive to sleep loss. A recent placebo controlled study at the DDRU (Stough et al. 2007) in which MDMA was administered to 60 participants showed performance was significantly worse when under the influence of MDMA. Drivers were more likely to signal incorrectly, undertake dangerous actions like skidding, brake inappropriately, follow cars at an unsafe distance and speed. While MDMA may improve some basic driving skills, it also impairs skills that are essential to safely drive in traffic, especially when combined with sleep loss.

Ketamine

Ketamine is primarily used as an anaesthetic agent and analgesic in medicine and veterinary practice (Mozayani, 2002). The hypnotic and hallucinogenic effects of this drug are gaining popularity among those who frequent dance parties as well as within the medical profession (Curran and Morgan, 2000; Dillon et al., 2002; Moore and Bostwick, 1999). Psychological effects include novel body sensations, dissociation, confusion and feelings of invulnerability (Mozayani, 2002). Physical effects include tachycardia, tachypnoea, blurred vision, uncoordination and catatonia (Dillon, 2002). Some users describe a spiritual near-death experience known as the K-hole (Mozayani, 2002). There have been no direct studies examining the effects of ketamine upon driving. Instead, studies have focused on subjective and objective data and correlate this to whether this may or may not impair driving. A study by Guillermo et al. (2001) revealed that reaction time is adversely affected by even subacute doses of ketamine. A study undertaken on dance party volunteers who tested positive to ketamine revealed that 90 per cent of subjects with ketamine above the cut-off concentrations of saliva 300 ng/mL were deemed drug-impaired as determined by roadside impairment tests (Cheng et al., 2006). Further research on ketamine and driving is required in a controlled clinical setting.

Opioids

Opioid drugs (of which heroin is one) are commonly used in the community. Not only are they prescribed for their analgesic and antitussive properties but their highly addictive potential has led them to be misused throughout the community.

Following an initial intense surge of euphoria, heroin users alternate between a wakeful and drowsy state known as 'the nod'. Other common effects include lethargy, disconnectedness, self-absorption, mental clouding, delirium, the inability to concentrate and diminished reflexes. Peak effects last for 1–2 hours, and the overall effects wear off in 3–5 hours, depending on dose. Due to ethical constraints there are very few experimental studies assessing heroin effects. An early study conducted in the 1960s found that heroin produced subjective feelings of sedation and slowed reaction (Martin and Fraser, 1961).

Methadone and buprenorphine have been found to be an effective long-term maintenance therapy for those addicted (Ogden and Moskowitz, 2004). Due to this prevalence, studies have been undertaken to review whether these substances adversely affect cognition and psychomotor performance on driving ability.

Whilst some studies reveal impairment (Darke et al., 2000; Specka, 2000; Zacny et al., 1994), other studies (Brooke et al., 1998; Hanks et al., 1995; O'Neill et al., 2000; Schindler et al., 2004) found that performance impairment does not differ significantly in comparison to drug free subjects. Results appear to be dose related with effects differing between opioid naïve and opioid dependent subjects. O'Neill et al. (2000), Hanks et al. (1995), and Brooke et al. (1998) administered 10 to 15 mg of morphine, either as an acute dose or every four hours and found no significant impairment in driving tasks. In fact morphine was found to improve reaction time. Nevertheless, whilst Zacny et al. (1994) found that hand-eye coordination was not affected by 2.5, 5 or 10 mg/70 kg doses of intravenous morphine, other psychomotor skills were impaired relative to the dose administered. Bruera et al. (1989) examined patients taking morphine-based drugs for pain relief. Those patients who received an increased dose within the previous three days experienced significant cognitive impairment, with no difference noted on those patients who were stabilized on a set dose. Chesher et al. (1989) found similar results when comparing driving ability between patients stabilized on methadone and those beginning or receiving an increased dose of methadone.

A key finding appears to be that patients on a maintenance dose have no psychomotor impairment in relation to driving ability. This is likely to be due to an increased opioid tolerance (Fishbain et al., 2003). Therefore patients are often recommended to cease driving for up to four weeks until their opioid regime is stabilized (Drug and Alcohol Services, 2006; Smith, 1996).

Benzodiazepines

Benzodiazepines are the most common psychotropic drugs prescribed in Australia and are frequently used for the treatment of anxiety and panic disorders. Although drugs classed in this category have varying effects on individuals, they are all known to have some degree of sedative, hypnotic and anxiolytic action. Benzodiazepines are also widely used in a sub-set of the community to reduce the adverse consequences of other drug use. For example, central nervous system depressants may be used to try to alleviate withdrawal or 'crash' symptoms that arise from heavy amphetamine use.

There is a considerable lack of conclusive research on the effects of benzodiazepines on driving skills, in particular in patient populations for which the drug is intended. To date, research on the effects of benzodiazepines on psychomotor performance have provided inconsistent results. The reports suggest that benzodiazepines can either impair (for example, O'Hanlon and Ramaekers, 1995) or have no effect on performance tasks that assess driving related skills (Hindmarch, 1995). Improvement in performance has even been reported (Wetherell, 1979) which suggests that the variation in findings may be, in part, due to the variation in an individual's response to the drug itself, withdrawal effects and whether the drug is abused or taken to improve functioning in various patient conditions.

Other Prescription Medications

There are many prescription drugs of which only a few classes and a limited range of actual drugs within those classes have been studied experimentally. The following prescription drugs that are reviewed below (and previously for example, benzodiazepines) are often reported in studies showing an impairment in driving behaviours, cognitive and motor skills or have been shown to increase crash risk in epidemiological studies.

Antidepressants

Depression can cause symptoms of poor concentration and attention disturbances as well as memory and executive functional deficits which may impair driving ability (Austin et al., 2001). The objective of pharmacological treatment for depression is to allow the patient to participate fully in the activities of daily living, and this includes driving (Brunnauer et al., 2006). Nevertheless, antidepressants may impair driving due to their effects on sedation, agitation and insomnia (Brunnauer et al., 2006). Studies have revealed that different classes of antidepressants have differing effects on driving ability. A study by Brunnauer et al. examined the effects of tricyclics (TCAs), selective serotonin reuptake inhibitors (SSRIs), noradrenergic specific serotonergics (NASSAs) and selective noradrenalin reuptake inhibitors (SNRIs) on driving ability and noted differences in the measures of reactivity, selective attention and stress tolerance. SSRIs were found to cause less impairment in psychomotor and cognition functioning (Brunnauer et al., 2006; Edwards, 1995) than TCAs (Brunnauer et al., 2006; Ray et al., 1992). Mirtazapine (NASSA) has a sedative effect due to its antihistamine effects. This is particularly apparent when taken as a morning dose (Brunnauer et al., 2006; Ridout et al., 2003). However, those who take mitrazipine as a nocturnal dose show improvement in their performance based tasks as opposed to untreated depressives (Brunnauer et al., 2008). The Brunnauer et al. (2008) study revealed that only 10 per cent of participants on TCAs passed the tests without impairments compared to 20 per cent on venlafaxine (SNRI), 28 per cent on an SSRI and 50 per cent on a NASSA.

Nevertheless, a review of the studies reveal mixed results (Randy et al., 2009). Some studies have found that TCAs have a greater relative risk for motor vehicle accidents (Ray et al., 1992; Leveilleet al., 2009), whilst others have found no increased risk (Barbone et al., 1998). This variance in findings could be due to the differing ages of participants as well as the variance in treatment periods (Randy et al., 2009). TCAs have antihistaminic and anticholinergic effects which may cause sedation (Randy et al., 2009). Most individuals adapt to these side effects after the initial treatment period, particularly if they are not elderly, have no evidence of hepatic dysfunction and are maintained on a stable dose (Randy et al., 2009). As there appears to be a potential risk, counselling regarding driving safety needs to be tailored to the individual, taking into account such factors as age, type and dose of antidepressant, phase of treatment, co-administered psychotropic drugs and individual insight into psychomotor and cognitive impairment (Randy et al., 2009; Brunnauer et al., 2006). Again the role of the clinical illness of depression or anxiety may cause significant impairment in driving. When most studies have been conducted in healthy controls it is difficult to ascertain the relative impairment of depression versus anti-depressant therapy. A depressed, un-medicated patient may be more at risk of accident than a medicated depressed patient, even though the medication may itself impair driving. Additionally because different classes of anti-depressants work differently on the brain it is difficult at this stage of research to be conclusive about the impairing effects of anti-depressants in general.

Antihypertensives

All antihypertensives can impact driving due to their vasomotor effects. In the initial stage of treatment, hypotension and dizziness may occur. Centrally acting antihypertensives, such as clonidine, methyldopa and reserpine may have a sedative action which calls for particular attention (Castot et al., 2009). A comparison study between methyldopa and reserpine and the beta blocker atenolol revealed that whilst driving performance deteriorated in those who were on the centrally acting antihypertensives, kinetic visual acuity (KVA) improved significantly for those on atenolol (Betts, 1981). Atenolol does cause dizziness, however, this is short in duration, moderate to weak in action and resolves with a lower dose (Almova and Elgarov, 1999). Angiotensin converting enzyme (ACE) inhibitors such as ramipril have side effects of dizziness, drowsiness, vertigo, hypotension, hyperkalemia and syncope which may impair driving (Almova and Elgarov, 1999; McGwin et al., 2000). Calcium channel blockers have been associated with a decreased risk of motor vehicle accidents. This could be due to their effects on stabilizing abnormal heart rhythms and relieving angina (McGwin et al., 2000). Patients need to be informed regarding the risks of taking these medications and driving, particularly in the treatment initiation stage (Castot et al., 2009). However, there are few studies in this area.

Oral Hypoglycaemics

Hypoglycaemia amongst those with Insulin or Non-Insulin Dependent Diabetes Mellitus (IDDM and NIDDM) may result in cognitive-motor slowing and loss of consciousness. This could negatively impact driving performance (Hemmerlgarn et al., 2006). This risk is more associated with mismanagement of dosage regimes, reduced food intake or strenuous exercise than with the medication itself (Castot et al., 2009).

A study by Hemmelgarn et al. (2006) explored whether antidiabetic medication increased the chances of a motor vehicle crash. They concluded that the use of insulin alone or a combination of sulfonylurea and metformin increased the risk of a MVC by 30 to 40 per cent, particularly for those on a higher dose. This risk is higher for those who take a combination of insulin and oral hypoglycaemics. On the contrary, no increased risk was found for those who are on oral monotherapy. Hemmelgarn et al. are unclear whether this small increased risk is medication induced or related to latter stage diabetic complications such as retinopathy and neuropathy, given that treatment regimes correlate with disease progression. However, a driving simulator study on a group of 25 IDDM patients found that whilst driving performance was not impaired with mild hypoglycaemia (3.6 mmol/l), moderate hypoglycaemia (2.6 mmol/l) reduced driving performance with more swerving, spinning, crossing the midline and compensatory slow driving (Cox et al., 2000). Blood glucose levels are advised to be tested prior to driving, especially in those who have asymptomatic hypoglycaemia (Hemmelgarn et al., 2006; Cox et al., 2000).

Antihistamines

Antihistamines are H1-receptor antagonists which are used widely by those suffering seasonal or chronic allergic rhinitis (Stork et al., 2006). Because of their widespread use, many studies (for example, O'Hanlon et al., 1995) have been undertaken to assess whether they impact driving. First generation antihistamines, such as brompheniramine, dimenhydrinate and diphenhydramine have lipophillic properties which enable cross over through the blood brain barrier. The depressive action on the central nervous system can cause drowsiness, dizziness, un-coordination and increased reaction time (Jauregi et al., 2006). The sedative effects persist into the next day (Kay, 2000). Anticholinergic effects of dry mouth and blurry vision are also evident (Jauregi et al., 2006).

Second generation antihistamines, such as loratadine, may also impair driving through psychomotor retardation and drowsiness (Castot et al., 2009; Verster and Edmund, 2004). However, the effects are generally milder than those of first generation antihistamines and depend on gender, dose and time of intake (Castot et al., 2009; Verster and Edmund, 2004). Tolerance tends to develop after four to five days (Verster and Edmund, 2004). O'Hanlon et al. (1995) found that whilst impairment is sometimes evident at higher than recommended doses, impairment is small to almost undetectable at recommended doses. One exception to this is cetirizine (Hennessy and Strom, 2000). Fexofenadine and levocetirizine are third-generation antihistamines which have been shown to produce no driving impairment (Verster and Edmund, 2004). Weiler et al. (2000) compared fexofenadine (60 mg), diphenhydramine (50 mg), alcohol (0.1 per cent blood alcohol concentration) and a placebo on driving performance. They found little difference in driving performance between those participants on fexofenadine or placebo. However, those participants in the diphenhydramine group performed worse on the driving tasks than those in the alcohol group.

Issues

There is still much unknown about the effect of many drugs on driving (particularly poly-drug use which is outside the scope of this chapter) although there is some consensus about the impairing effects of drugs such as cannabis and ketamine which significantly slow CNS activity. There is a lack of information and research on driving on many prescription drugs (such as modafinil). There is also a lack of knowledge about how low levels of alcohol (in other words less than .05) combined with various drugs (licit and illicit) may impair driving. This information needs to be made public if public behaviours are to be changed. There is a lack of understanding of whether some drugs improve driving skills in certain types of patients who use the drugs based on prescriptions. For instance does a patient with anxiety disorders drive better after taking a prescribed medication than without medication to treat his/her disorder? Understanding whether patients with various drug and alcohol addictions should be allowed licences and how potential future accidents in these drivers can be avoided is important. Is it possible to suspend licences in patients who have an addiction before they have had an accident? On what basis can a previously addicted person prove that they are no longer addicted to alcohol or a drug that impairs driving? In this respect hair analysis should be investigated as a tool in identifying potential drivers who are still dependent on drugs. This could then be used to prevent currently addicted drivers from driving.

Perhaps one of the biggest problems is in the area of truck driving. Trucks accidents have a disproportionate contribution to the road toll in Australia, relative to motor cars. Saliva testing has identified the use of amphetamines in truck drivers who use these substances to drive further and for longer durations. Unless there is regular drug-detection for drivers then this issue will remain a problem as there is an obvious economic advantage for truck drivers (and their employers) to find ways in which they can drive a truck further distances.

References

Arnedt, J.T., Wilde, G.J.S., Munt, P.W., and MacLean, A.W. (2000). Simulated driving performance following prolonged wakefulness and alcohol consumption: Separate and combined contributions to impairment. *Journal of Sleep Research, 9*, 233–41.

Almova, I., and Elgarov, I. (2000). Atenolol and ramipril in arterially hypertensioned motor vehicle drivers: efficacy and safety. Proceedings of *T2000 – 15th conference on alcohol, drugs and traffic safety*, Stockholm, Sweden.

Arnedt, J.T., Wilde, G.J.S., Munt, P.W., and MacLean, A.W. (2001). How do prolonged wakefulness and alcohol compare in the decrements they produce on a simulated driving task? *Accident Analysis and Prevention, 33*, 337–44.

Austin, M., Mitchell, P., and Goodwin, G. (2001). Cognitive deficits in depression: Possible implications for functional neuropathy. *British Journal of Psychiatry, 178*, 200–206.

Australian Institute of Health and Welfare (2008). *2007 National Drug Strategy Household Survey: First results.* Drug Statistics Series number 20. Cat. no. PHE 98. Canberra: AIHW.

Australian Transport Safety Bureau (2007). *Road Deaths Australia 2006 Statistical Summary.* ATSB Research and Analysis Report, Road Safety, Australian Government.

Barbone, F., McMahon, A., and Davey, P. (1998). Use of benzodiazepines increased road traffic accidents whereas use of tricyclic antidepressants and SSRIs did not. *Lancet, 24.* 1331–36

Berghaus, G., Scheer, N., and Schmidt, P. (1995). Effects of Cannabis on Psychomotor Skills and Driving Performance - a Meta-analysis of Experimental Studies. In C.N. Kloeden, and A.J. Mclean (Eds.), *Proceedings – 13th International Conference on Alcohol, Drugs and Traffic Safety.* NHMRC Road Accident Research Unit, The University of Adelaide, pp. 403–409.

Berry, J.G., and Harrison, J.E. (2008). *Serious Injury Due to Land Transport Accidents Australia 2005–06.* Australian Institute of Health and Welfare. Injury research and statistics series number 42. Cat. no.INJCAT 113. Adelaide: AIHW.

Betts, T. (1981). Effects of beta blockade on driving. *Aviation, Space and Environmental Medicine, 11*(2), 540–45.

Brooke, C., Ebnhage, A., Fransson, B., Haggi, F., Jonzon, B., Kraft, I., and Wesnes, K. (1998). The effects of intravenous morphine on cognitive function in healthy volunteers. *Journal of Psychopharmacology, 12*, A45.

Bruera, E., MacMillan, K., Hanson, J., and MacDonald, R. (1989). The cognitive effects of the administration of narcotic analgesics in patients with cancer pain. *Pain, 39*, 13–16.

Brunnauer, A., Laux, G., David, I., Fric, M., Hermisson, I., and Moller, H. (2008). The impact of reboxitine and mirtazapine on driving simulator performance and psychomotor function in depressed patients. *Journal of Clinical Psychiatry, 69*, 1800–1806.

Brunnauer, A., Laux, G., Geiger, E., Soyka, M., and Moller, H. (2006). Antidepressants and driving ability: Results from a clinical study. *Journal of Clinical Psychiatry, 67*, 1176–1781.

Castot, A., Delorme, B., and Tricotel, A. (2009). *Medicinal products and driving.* Paris: AFSSAPS, pp. 1–26

Cheng, W., Ng, K., Chan, K., Mok, V., and Cheung, B. (2006). Roadside detection of impairment under the influence of ketamine: Evaluation of ketamine impairment symptoms with reference to its concentration in oral fluid and urine. *Forensic Science International, 170*, 51–8.

Ch'ng, C.W., Fitzgerald, M., Gerostamoulos, J., Cameron, P., Bui, D., Drummer, O.H., Potter, J., and Odell, M. (2007). Drug use in motor vehicle drivers presenting to an Australian, adult major trauma centre. *Emergency Medicine Australasia, 19*, 359–65.

Chesher, G., Lemon, J., Gomel, M., and Murphy, G. (1989). *The effects of methadone, as used in a methadone maintenance program, on driving related skills*. Sydney: NDARC.

Cox, D., Gonder-Frederick, L., Kovatchev, B., Julian, D., and Clarke, W. (2000). Progressive hypoglycaemia's impact on driving simulation performance. *Diabetes Care, 23*, 163–70.

Curran, H., and Morgan, C. (2000). Cognitive, dissociative and psychogenic effects of ketamine in recreational users on the night of drug use and 3 days later. *Addiction, 95*, 575–90.

Darke, S., Sims, J., MacDonald, S., and Wickes, W. (2000). Cognitive impairment among methadone patients. *Addiction, 95*, 687–95.

Dawson, A., and Reid, K. (1997). Fatigue, alcohol and performance impairment. *Nature, 388*, 235.

De Waard, D., Brookhuis, K.A., and Pernot, L.M.C. (2000). A driving simulator study on the effects of MDMA (Ecstasy) on driving performance and traffic safety. *Proceedings of the International Council on Alcohol Drugs and Traffic Safety* (ICADTS), Stockholm (Sweden), May 2000.

Dillon, P., Copeland, J., and Jansen, K. (2002). Patterns of use and harms associated with non-medical ketamine use. *Drug and Alcohol Dependence, 69*, 23–38.

Drug and Alcohol Services (2006). *Benzodiazepines, opioids and driving prescriber resource kit*. Adelaide: Drug and Alcohol Services, Government of South Australia.

Drugs and Crime Prevention Committee (2004). *Inquiry into Amphetamine and 'Party Drug' Use in Victoria – Final Report*. DCPC, Parliament of Victoria.

Drummer, O.H., Gerostamoulos, J., Batziris, H., Chu, M., Caplehorn, J.R.M., Robertson, M.D., et al. (2003). The incidence of drugs in drivers killed in Australian road traffic crashes. *Forensic Science International, 134*(2–3), 154–62.

Drummer, O.H., Gerostamoulos, J., Batziris, H., Chu, M., Caplehorn, J., Robertson, M.D. et al. (2004). The involvement of drugs in drivers of motor vehicles killed in Australian road traffic crashes. *Accident Analysis and Prevention, 36*, 239–48.

Drummer, O.H. (1994). *Drugs in drivers killed in Victorian road traffic accidents. The use of responsibility analysis to investigate the contribution of drugs to fatal accidents*. Victorian Institute of Forensic Pathology and Monash University Department of Forensic Medicine: Report No. 0394.

Drummer, O.H. (1998). *Drugs in drivers killed in Victorian road traffic accidents*. Melbourne: Victorian Institute of Forensic Medicine: Report No. 0298.

Edwards, J. (1995). Drug choice in depression: Selective serotonin reuptake inhibitors or tricyclic antidepressants? *CNS drugs, 4*, 141–59.

Falleti, M.G., Maruff, P., Collie, A., Darby, D.G., and McStephen, M. (2003). Qualitative similarities in cognitive impairment associated with 24 h of sustained wakefulness and a blood alcohol concentration of 0.05%. *Journal of Sleep Research, 12*, 265–74.

Fishbain, D., Cutler, R., and Rosomoff, H. (2003). Are opioid-dependent/tolerant patients impaired in driving-related skills? A structured evidence-based review. *Journal of Pain and Symptom Management, 25*, 559–77.

Guillermo, Y., Micallef, J., Possamai, C., Blin, O., and Hasbroucq, T. (2001). N-methyl-D-aspartate receptors and information processing: Human choice reaction time under a subanaesthetic dose of ketamine. *Neuroscience Letters, 303*, 29–32.

Hanks, G., O'Neill, W., Simpson, P., and Wesnes, K. (1995). The cognitive and psychomotor effects of opioid analgesics. II. A randomised controlled trial of single doses of morphine, lorazepam and placebo in healthy subjects. *European Journal of Clinical Pharmacology, 48*, 455–60.

Hemmelgarn, B., Levesque, L., and Suissa, S. (2006). Anti-diabetic drug use and the risk of motor vehicle crash in the elderly. *Canadian Journal of Clinical Pharmacology, 13*, 112–20.

Hennessy, S., and Strom, B. (2000). Non-sedating antihistamines should be preferred over sedating histamines in patients who drive. *Annals of Internal Medicine, 132*, 405–407.

Hindmarch, I. (1995). The Psychopharmacology of Clobazam. *Human Psychopharmacology: Clinical and Experimental, 10*, 15–25.

Jackson, P.G., Tunbridge, R.J., and Rowe, D.J. (2000). Drug recognition and field impairment testing: Evaluation of trials. In *Alcohol, Drugs and Traffic Safety: Proceedings of the 15th International Conference on Alcohol, Drugs and Traffic Safety*, May 21–26, 2000, Stockholm, Sweden.

Jauregi, I., Mullol, J., Barta, J., Cuvillo, A., Davila, I., Montoro, J., Sastre, J., and Valero, A. (2006). H1 antihistamines: Psychomotor performance and driving. *Journal of Investigational Allergology and Clinical Immunology, 161*, 37–44.

Kay, G. (2000). The effects of antihistamines on cognition and performance. *Journal of Allergy and Clinical Immunology, 105*, 622–7.

Kelly, E., Darke, S., Ross, J., and Kelly, E. (2004). A review of drug use and driving: Epidemiology, impairment, risk factors and risk perceptions. *Drug and Alcohol Review, 23*, 319–44.

Kuypers, K., Samyn, N., and Ramaekers, J. (2006). MDMA and alcohol effects, combined and alone, on objective and subjective measures of actual driving performance and psychomotor. *Psychopharmacology, 187*, 467–75.

Kuypers, K.P.C., Wingen, M., Samyn, N., Limbert, N., and Ramaekers, J.G. (2007). Acute effects of nocturnal doses of MDMA on measures of impulsivity and psychomotor performance throughout the night. *Psychopharmacology, 192*(1), 111–19

Lamers, C.T., Ramaekers, J.G., Muntjewerff, N.D., Sikkema, K.L., Samyn, N., Read, N.L., Brookhuis, K. A., and Riedel, W. J. (2003). Dissociable effects of a single dose of ecstasy (MDMA) on psychomotor skills and attentional performance. *Journal of Psychopharmacology, 17*, 379–87.

Leveille, S., Buchner, D., and Koepsell, T. (1994). Psychoactive medications and injurious motor vehicle crashes in elderly drivers. *American Journal of Epidemiology, 5,* 591–8.

Logan, B.K. (1996). Methamphetamine and driving impairment. *Journal of Forensic Sciences, 41,* 457–64.

Longo, M.C., Hunter, C.E., Lokan, R.J., White, J.M., and White, M.A. (2000). The prevalence of alcohol, cannabinoids, benzodiazepines and stimulants amongst injured drivers and their role in driver culpability. Part II: The relationship between drug prevalence and drug concentration, and driver culpability. *Accident Analysis and Prevention, 32,* 623–32.

Martin, W.R., and Fraser, H.F. (1961). A study of physiological and subjective effects of heroin and morphine administered intravenously in post-addicts. *The Journal of Pharmacology and Experimental Therapeutics, 133,* 388–99.

McGwin, G., Sims, R., Pulley, L., and Roseman, J. (2000). Relations among chronic medical conditions, medications, and automobile crashes in elderly: A population-based case-control study. *American Journal of Epidemiology, 152,* 424–31.

Moore, N., and Bostwick, J. (1999). Ketamine dependence in anesthesia providers. *Psychosomatics, 40,* 356–9.

Mozayani, A. (2002). Ketamine-effects on human performance and behaviour. *Forensic Science Review, 14,* 124–31.

Mura, P., Chatelain, C., Dumestre, V., Gaulier, J.M., Ghysel, M.H., Lacriox, C., Kergueris, M.F., Lhermitte, M., Moulsma, M., Pepin, G., Vincent, F., and Kintz, P. (2006). Use of drugs of abuse in less than 30 year old drivers killed in a road crash in France: A spectacular increase for cannabis, cocaine and amphetamines. *Forensic Science International, 160,* 168–72.

Neutal, I. (1998). Benzodiazepine-related traffic accidents in young and elderly drivers. *Human Psychopharmacology: Clinical and Experimental, 13,* 115–23.

O'Hanlon, J., and Ramaekers, J. (1995). Antihistamine effects on actual driving performance in a standard test: A summary of Dutch experience. *Allergy, 50,* 234–42.

O'Hanlon, J., Vermeeren, A., Uiterwijk, A.A., van Veggel, L.M., and Swijgman, H.F. (1995). Anxiolytics' effects on the actual driving performance of patients and healthy volunteers in a standardized test: An integration of three studies. *Neuropsychobiology 31,* 81.

O'Neill, W., Hanks, G., Simpson, P., Fallon, M., Jenkins, E., and Wesnes, K. (2000). The cognitive and psychomotor effects of morphine in healthy subjects: A randomized controlled trial of repeated (four) oral doses of dextropropoxyphene, morphine, lorazepam and placebo. *Pain, 85,* 209–15.

Ogden, E., and Moskowitz, H. (2004). Effects of alcohol and other drugs on driver performance. *Traffic Control Prevention, 5,* 185–98.

Ogden, E., Frederiksen, T., Stough, C., and King, R. (2009). Responsibility Analysis: Screening of drugs in blood samples from injured drivers in Victoria.

Poster presented at The Australasian Professional Society on Alcohol and other Drugs, Darwin, Australia 1–4 November.

Papafotiou, K., Carter, J.D., and Stough, C. (2005). The relationship between performance on the standardised field sobriety tests, driving performance and the level of Î"9-tetrahydrocannabinol (THC) in blood. *Forensic Science International, 155*(2–3), 172–8.

Parrott, A.C., and Lasky, J. (1998). Ecstasy (MDMA) effects upon mood and cognition: Before, during and after a Saturday night dance. *Psychopharmacology, 139* (3), 261–8.

Ramaekers, J., and Kuypers, K. (2006). Acute effects of 3,4-methylenedioxymethamphetamine (MDMA) on behavioral measures of impulsivity: Alone and in combination with alcohol. *Neuropsychopharmacology, 31*, 1048–55.

Ramaekers, J.G., Berghaus, G., van Laar, M., and Drummer, O.H. (2004). Dose related risk of motor vehicle crashes after cannabis use. *Drug and Alcohol Dependency, 73*(2), 109–19.

Ramaekers, J.G., Kuypers, K.P.K., and Samyn, N. (2006). Stimulant effects of 3,4-methylenedioxymethamphetamine (MDMA) 75 mg and methylphenidate 20 mg on actual driving during intoxication and withdrawal. *Addiction, 101*, 1614–21.

Ramaekers, J.G., Robbe, H.W., and O'Hanlon, J.F. (2000). Marijuana, alcohol and actual driving performance. *Human Psychopharmacology, 15*(7), 551–8.

Ramaekers, J., Uiterwijk, M., and O'Hanlon, J. (1992). Effects of loratadine and cetirizine on actual driving and psychometric test performance and EEG during driving. *European Journal of Clinical Pharmacology, 42*, 363–9.

Randy, A., Sansone, M., and Lori, A. (2009). Driving on antidepressants: Cruising for a crash? *Psychiatry, 6* (9), 13–16.

Ray, W., Fought, R., and Decker, M., (1992). Psychoactive drugs and the risk of injurious motor vehicle crashes in elderly drivers. *American Journal of Epidemiology, 136*, 873–83.

Ridout, F., Meadows, R., and Johnsen, S. (2003). A placebo controlled investigation into the effects of paroxetine and mirtazapine on measures related to car driving performance. *Human Psychopharmacology, 18*, 261–9.

Robertson, M.D., and Drummer, O.H. (1994). Responsibility analysis: A methodology to study the effects of drugs in driving. *Accident Analysis and Prevention, 26*, 243–7.

Schindler, S., Ortner, R., Peternell, A., Eder, H., Opgenoorth, E., and Fischer, G. (2004). Maintenance therapy with synthetic opioids and driving aptitude. *European Addiction Research, 10*, 80–87.

Schwilke, E.W., Sampaio dos Santos, M.I., and Logan, B.K. (2006) Changing patterns of drug and alcohol use in fatally injured drivers in Washington state. *Journal of Forensic Science, 51*, 1191–98.

Sexton, B.F., Tunbridge, R.J., Brooke-Carter, N., Jackson, P.G., Wright, K., Stark, M.M., and Englehart, K. (2000). *The Influence of Cannabis on Driving.*

Technical Report 477, United Kingdom: Transport Research Foundation Limited (TRL).

Silber, B.Y., Croft, R.J., Papafotiou, K., and Stough, C. (2006). The acute effects of d-amphetamine and methamphetamine on attention and psychomotor performance. *Psychopharmacology, 187*(2), 154–69.

Silber, B.Y., Papafotiou, K., Croft, R.J., Ogden, E., Swann, P., and Stough, C. (2005). The effects of dexamphetamine on simulated driving performance. *Psychopharmacology, 179*(3), 536–43.

Smiley, A. (1986). Marijuana: On-road and driving simulator studies. *Alcohol, Drugs and Driving, 2* (3–4), 121–34.

Smiley, A. (1998). *Marijuana: On road and driving simulator studies*. Prepared for World Health Organization, Geneva, Switzerland.

Smith, A. (1996). Patients taking stable doses of morphine may drive. *British Medical Journal, 312*(7022), 56–7.

Specka, M., Finkbeier, T., Lodemann, E., Leifert, K., Kluwig, J., and Gastpar, M. (2000). Cognitive-motor performance of methadone-maintained patients. *European Addiction Research, 6*, 8–19.

Stein, A., Allen, R., Cook, M., and Karl, R. (1983). *A Simulator Study of the Combined Effects of Alcohol and Marijuana on Driving Behavior- Phase II*. Report DOT HS-5–01257. Washington, DC, National Highway Traffic Safety Administration.

Stork, A., Haeften, T., and Veneman, T. (2006). Diabetes and driving, desired research methods and their pitfalls, current knowledge, and future research. *Diabetes Care, 29*(8), 1942–9.

Stough, C., Ogden, E. and Papafotiou, K. (2007). *Roadside saliva based testing for amphetamine-type stimulants in drivers: An evaluation of the relationship between positive drug tests and driving impairment after the consumption of methamphetamine and MDMA*. Australian Research Council Discovery Grant.

Traffic Accident Commission. (2009). Victoria, Australia www.tacsafety.com.au / jsp/ homepage/home.jsp

Verster, J., and Edmund, R. (2004). Antihistamines and driving ability: Evidence from on-the-road driving studies during normal traffic. *Annals of Allergy, Asthma and Immunology, 92*(3), 294–30.

Weiler, J., Bloomfield, J., Woodworth, G., Grant, A., Layton, T., Brown, T., McKenzie, D., Baker, T., and Watson, G. (2000). Effects of fexofenadine, diphenhydramine, and alcohol on driving performance. A randomized, placebo-controlled trial in the Iowa driving simulator. *Annals of Internal Medicine, 132*, 354–63.

Wetherell, A. (1979). Individual and group effects of 10 mg diazepam on drivers' ability, confidence and willingness to act in a gap-judging task. *Psychopharmacology, 63*(3), 259–67.

Williamson, A.M., and Feyer, A-M. (2000). Moderate sleep deprivation produces comprehensive cognitive and motor performance impairments equivalent to legally prescribed levels of alcohol intoxication. *Occupational and Environmental Medicine, 57*, 649–55.

Zacny, J., Lichtor, J., Fleming, D., Coalson, D., and Thompson, W. (1994). A dose-response analysis of the subjective, psychomotor and physiological effects of intravenous morphine in healthy volunteers. *Journal of Pharmacology and Experimental Therapy, 268*, 1–9.

Chapter 18

Effects of Snowfall on Seat-Belt Use

Özlem Şimşekoğlu and Timo Lajunen

Department of Psychology, Izmir University of Economics, Turkey

Introduction

Many studies have demonstrated that adverse weather conditions such as rainfall and snowfall increase traffic accident risk considerably (for example, Andrey et al., 2003). For instance, compared to rainfall snowfall has a greater effect on collisions, and the accident risk is highest for the first snowfall of the season, according to the literature review by Andrey et al. (2003). Similarly, the first snowfall days of the year were found to have increased traffic accident rates including fatal accidents (Eisenberg and Warner, 2005).

As well as traffic accident risk, adverse weather conditions also affect driver behaviour (Anttila et al., 1999; Kilpeläinen and Summala, 2007). Drivers seem to acknowledge, to some degree, the heightened risk caused by bad weather and road conditions. For instance, Edwards (1999) observed vehicle speeds in fine, rainy and misty weather conditions and found a small but statistically significant reduction in mean speed in wet weather and misty conditions. However, Edwards (1999) concluded that this reduction in speeds was not sufficient to compensate for the increased accident risk caused by the adverse weather conditions. Similarly, studies conducted in Nordic countries (Finland, Norway and Sweden) show that drivers using studded tyres drive faster in the curves than those with non-studded tyres, while still maintaining a somewhat larger safety margin (Rumar et al., 1976). Hence, drivers seem to take bad weather conditions into account, although their compensative behaviours (for example speed reduction, larger safety margins) seem to be insufficient to compensate for the negative effects of the weather.

Almost all studies about weather conditions and driver behaviour have concentrated on driver behaviours like speed choice, following distance and maintenance of safety margins in general. Very few studies have investigated the effects of bad weather conditions on the use of protective equipment like seat-belts, and most of these studies have involved surveys about factors affecting self-reported seat-belt use (for example, Chliaoutakis et al., 2000). Also, most of the studies about the effects of adverse weather conditions on driving have been conducted in Nordic countries (Finland, Norway, Sweden) or in North America, where local road administrations normally try to reduce the effect of bad weather conditions using some maintenance operations (for example, ploughing, salting, gritting and providing warnings using variable message signs and radio messages). Also, the use of winter tyres in these countries is often obligatory. However, in the Eastern Mediterranean countries, such as Turkey, much less support is provided by the local road administration to reduce the impact of the adverse weather conditions on driving.

Aims of the Study

The aim of the present study was to investigate the effects of adverse weather conditions (snowfall) on the frequency of seat-belt use and child restraint use among Turkish car occupants.

Method

Participants

The participants included a total of 611 observed front-seat occupants (383 drivers, 228 front seat passengers) and 137 children. The sample characteristics of all observed car occupants are displayed in Table 18.1.

Table 18.1   Sample characteristics

	First observation		Second observation	
	Driver	**Passengers**	**Driver**	**Passengers**
Sex				
Male	149 (81.4%)	32 (29.1%)	174 (87.0%)	27 (22.9%)
Female	34 (18.6%)	78 (70.9%)	26 (13.0%)	91 (77.1%)
Age (years): Adults				
< 30	45 (24.7%)	34 (31.2%)	13 (6.5%)	13 (11.0%)
30–50	128 (70.3%)	71 (65.1%)	178 (89.0%)	101 (85.6%)
> 50	9 (4.9%)	4 (3.7%)	9 (4.5%)	4 (3.4%)
Age (years): Children				
0–1	-	2 (1.1%)	-	-
2–3	-	11 (6.0%)	-	2 (2.8%)
4–7	-	12 (6.6%)	-	29 (40.8%)
8–10	-	26 (14.2%)	-	31 (43.7%)
11–13	-	14 (7.7%)	-	9 (12.7%)

Instrument and Data Collection

The present study included two data sets collected at two different times (in sunny dry weather and cold weather with snow on the road). In the two observation sessions, seat-belt use (yes, no), estimated age group (< 30 years, 30–50 years, > 50 years) and sex (male, female) of the observed drivers and front-seat passengers were recorded by each observer. Seating location (front seat, back seat, child seat) and age category (0–1 years, 2–3 years, 4–7 years, 8–10 years, 11–13 years) of the observed child passengers were also recorded by the observers. When there was more than one child passenger in the car, only one of the children (chosen at random) was recorded by the two observers. The sex of the children was not recorded.

Both of the observations were conducted in the car park of a shopping centre in Ankara, the capital city of Turkey, at two different time periods. This particular shopping centre was chosen because it had a large parking area with a high traffic flow. All observation sessions were made during the day (12.00–17.00) at the weekend. Before the observations, the observers were trained by the authors on how to make the observations. Two observers in each group made independent observations of car occupants, who may have consisted of only the driver or different combinations of driver, front-seat passenger and child passenger(s) in the car. Children sitting in the front seat of the car were not recorded as front seat passengers, but as child passengers. Observed cars were chosen randomly from the traffic flow entering the parking area of the shopping centre. There were a total of two observation sessions and each observation session lasted about two hours. Observation procedures were the same for the two observation periods except that the first observation was conducted in sunny dry weather and the second observation was conducted in cold weather with snow on the road.

Results

To test whether there was a significant difference in seat-belt use in the two weather conditions, separate Pearson χ^2 values were calculated. Percentages of seat-belt users within the two weather conditions are displayed in Table 18.2. Chi-squared analysis revealed that a significantly higher proportion of the observed drivers, front-seat passengers and child passengers were using a seat-belt in bad weather than in good weather conditions. Also, Pearson χ^2 values were calculated to see whether seating position of the child passengers changed significantly between the two weather conditions. Table 18.3 shows that there was no significant change in seating positions of the child passengers between the two weather conditions.

Table 18.2   Percentages of seat-belt users in the two weather conditions

	Good weather	Bad weather	Pearson $\chi^2_{(1)}$
Driver	16.9	30.0	9.0**
N	31	60	
Men	15.4	27.0	6.3**
N	23	47	
Women	23.5	50.0	4.5*
N	8	13	
Front seat passenger	14.5	30.5	8.2**
N	16	36	
Men	9.4	14.8	0.4
N	3	4	
Women	16.7	35.2	7.4**
N	13	32	
Child passenger	3.0	12.7	4.3*
N	2	9	

*$p < 0.05$; **$p < 0.01$*

Table 18.3   Children's seating positions in the two weather conditions (%)

	Good weather	Bad weather
Seating position		
Front seat	19.7	21.1
N	13	15
Back seat	74.2	73.2
N	49	52
Child seat	6.1	5.6
N	4	4

Pearson $\chi^2_{(3)} = 0.1$, $p = $ n.s

Discussion

The results of the present study showed that the drivers and passengers used a seat-belt more often in adverse weather conditions than in normal weather conditions, which may be interpreted as a compensation mechanism for the effects of the adverse weather conditions. This finding is in line with some previous survey findings indicating that travelling in bad weather conditions was positively related to seat-belt use (Chliaoutakis et al., 2000). It should be noted, however, that except for the change in the weather there may be other reasons for the increase in seat-belt use, such as changes in sample characteristics between the two observations. For example, compared to the sample in the first observation period, the sample in the second observation period included a considerably smaller number of young front-seat occupants. Considering that previous research has reported lower seat-belt use rates in younger people (for example, Jonah, 1990) the higher proportion of middle-aged front-seat occupants in the second observation period may partly explain the higher rate of seat-belt use found in the second observation period.

The fact that the seat-belt use of front seat occupants increased in adverse weather condition gives some idea about the nature of seat-belt use among Turkish car occupants: seat-belt use is not an automatic habit, but the decision is likely to be made each time based on situational factors such as the weather. The problem with this type of decision-making process is that it may be based on incorrect beliefs about the effectiveness of seat-belt use and inaccurate assessment of accident risk. This type of decision-making process regarding seat-belt use is also likely to be the reason for the low seat-belt use rates and the heterogeneity in the risk awareness about not using a seat-belt in the general Turkish population.

The weather effect on 'passive' (meaning that the aim is to reduce injury risk in the case of an accident) protection is especially interesting, because it reflects mostly the drivers' risk evaluation, whereas 'active' (meaning that the aim is to lessen the likelihood of an accident by, for example, reducing speed) compensatory behaviours are based both on risk assessments and direct feedback from the vehicle and the road. In other words, drivers might drive slower in bad weather because they assess accident risk to be high and/or because they feel that higher speed would result in loss of control of the vehicle. Hence, the decrease in speed in bad weather conditions may be based on both the overall assessment of risk and the feedback from the vehicle. The increase in seat-belt use from 16.9 to 30.0 per cent among drivers and from 14.5 to 30.5 per cent among front-seat passengers shows that car occupants assessed accident risk to be higher in bad weather than in good weather, but the increase was far from sufficient in terms of safety – only about one third of the drivers and passengers used a seat-belt even in the bad weather conditions.

In order to increase seat-belt use among Turkish car occupants, seat-belt use interventions should aim at habit formation rather than only emphasizing the rational and correct decision-making process. In the process of habit formation, strict police enforcement of seat-belt use combined with information campaigns are essential. Currently, in Turkey the police enforcement for controlling the use of seat-belts is very low and; therefore, the decision of whether to use a seat-belt/child restraint is almost completely left up to the individual drivers or passengers. However, enhanced enforcement by the police and carefully planned campaigns involving all parties (for example, mass media, schools, driving schools, kindergartens) are essential for the development of seat-belt use as a habit.

The present study has several limitations. Firstly, the sample of the present study may not be representative of the general population of Turkish drivers and passengers because the data were only collected in the car park of one shopping centre in Ankara. Another limitation of the study is that there was a time lag between the two observation sessions and some other time-related factors, apart from the weather, could have influenced seat-belt use and the driver sample. However, since weather conditions are an uncontrollable natural event, which cannot be manipulated to match the demands of study design, we did not have any other option but to have a time lag between the two observation sessions while waiting for the snowfall.

Acknowledgements

This work has been supported by EU Marie Curie Transfer of Knowledge programme ('SAFEAST' Project No: MTKD-CT-2004–509813) and the Graduate School of Psychology in Finland.

References

Andrey, J.M.B., Mills, B., Leahy, M., and Suggett, J. (2003). Weather as a Chronic Hazard for Road Transportation in Canadian Cities. *Natural Hazards, 28,* 319–43.

Anttila, V., Nygård, M., and Rämä, P. (2001). *Liikennesää-tiedotuksen Toteutuminen ja Arviointi Talvikaudella 1999–2000* (Helsinki, Finland: Tiehallinto. Liikenteen palvelut).

Chliaoutakis, E.J., Gnardellis, C., Drakou, I., Darviri, C., and Sboukis, V. (2000). Modelling the Factors Related to the Seatbelt Use by the Young Drivers of Athens. *Accident Analysis and Prevention, 32,* 815–25.

Edwards, J.B. (1999). Speed Adjustment of Motorway Commuter Traffic to Inclement Weather. *Transportation Research Part F, 2,* 1–14.

Eisenberg, D., and Warner, K.E. (2005). Effects of Snowfalls on Motor Vehicle Collisions, Injuries and Fatalities. *American Journal of Public Health, 95,* 120–24.

Jonah, B.A. (1990). Age Differences in Risky Driving. *Health Education Research, 5,* 139–49.

Kilpeläinen, M., and Summala, H. (2007). Effects of Weather and Weather Forecasts on Driver Behavior. *Transportation Research Part F, 10*, 288–99.

Rumar, K., Berggrund, U., Jernberg, P., and Ytterbom, U. (1976). Driver Reaction to a Technical Safety Measure: Studded Tires. *Human Factors, 18*, 443–54.

Chapter 19

Differences in Driving Behaviours between Elderly Drivers and Middle-Aged Drivers at Intersections

Nozomi Renge
Osaka University and SPS. Research Fellow

Masahiro Tada
Advanced Telecommunications Research Institute International

Kazumi Renge
Tezukayama University

Shinnosuke Usui
Osaka University

Introduction

The number of traffic fatalities in which elderly car drivers (65 years old and over) are at fault remains constant even though the number of such accidents for all drivers, regardless of age, has decreased in Japan. Elderly drivers cause many collisions at unsignalised intersections (Ikeda et al., 2004), small and mid-sized intersections (ITARDA, 2000), and intersections with stop signs (Oxley et al., 2006). Most of these accidents are head-on collisions and right-turn accidents (cars drive on the left in Japan). However, surprisingly few researchers have investigated elderly driver performance in residential areas. Therefore, this research focuses on the driving behaviour of elderly drivers at these types of risky intersections.

Previous research has shown that elderly drivers perform worse at controlling speed and searching, which significantly impacts upon other road users (Okamura and Fujita, 1997). Previous research has also shown that elderly drivers generally tend to look less in both directions at intersections (Renge et al., 2003). There are a number of systems for observing driving behaviour, including video cameras, driving simulators and eye trackers. Each system has its strengths and weaknesses, but there are also particular problems in dealing with visual search behaviours. To solve these problems, a technique to estimate visual search behaviours from a driver's head movement using small and lightweight wireless gyro head-mounted sensors were developed (Tada et al., 2009). Compared with other techniques, this method is able to record behaviour by attaching sensors to a driver and their car without increasing the driver's burden. The validity of this technique was demonstrated by the relationship between the sensors and an eye-tracking device (Tada et al., 2009). This technique is thus useful for observing driving behaviour for long periods while driving on actual roads.

The objective of the present study was to investigate the driving behaviour of elderly drivers at small unsignalised intersections with stop signs, where they cause many accidents. We compared driving behaviour between elderly drivers and middle-aged drivers, when going straight at intersections with and without stop signs. In particular, we focused on the change of driving speed and visual search behaviours for both the right and left directions.

Method

Participants

Field experiments were conducted with 56 drivers who were recruited from local residents. Nine participants who had been stopped at intersections with a stop sign by a flagman, due to road construction, were later excluded. Therefore, we analysed data from the remaining 27 healthy elderly drivers and 20 healthy middle-aged drivers. The healthy elderly drivers were all men, ranging in age from 63 to 78 years old ($M = 67.85$, $SD = 2.88$). The middle-aged drivers included two women and ranged in age from 30 to 57 years old ($M = 40.60$, $SD = 9.12$). Informed consent was obtained from all participants.

Materials

Two cars (Toyota Corolla, automatic transmission, 1500 cc) with CCD cameras to record the foreground, driver's head and foot position were included, along with the wireless 3D-gyro sensors attached to the drivers head. A DVR and SRcomm (M17, M64 – Datatec) and video cameras (Sony VX2000) recorded traffic scenes and driver behaviour. The gyro sensors attached to drivers (WAA-006 – Wireless Technology Inc., Advanced Telecommunications Research Institute International) recorded drivers' visual search behaviours and pedal operation, while the drivers drove a predetermined route (Figure 19.1). Another gyro sensor was attached to the dashboard to record the vehicles response. Moreover, we estimated driving speeds from the change in position per second using GPS. The measurement error using a normal GPS unit is ±5 m. However, the measurement error in this study was about ±1 m, as we performed matching processing to reduce the measurement error.

Figure 19.1 Gyro sensors and a driver with gyro sensors attached

Measures

Firstly, we measured the points at which the drivers' decelerated and accelerated and how driving speed changed when passing intersections. In order to identify how long in advance drivers recognise risks and prepare for them before approaching an intersection, the distance from the point where drivers started to decelerate to the centre of the intersection was measured (Figure 19.2). For example, -50 m means that the driver started to reduce his/her speed 50 m before the centre of the intersection. We also measured the distance from the point where drivers started to accelerate to the centre of an intersection. In order to indicate driving speed change, we chose driving speeds at three points, first, the point where drivers started to reduce their speed (the normal driving speed before drivers approached an intersection), second, a point 30 m before the centre of the intersection, and, third, the point where drivers started to accelerate (from the lowest speed in an intersection).

Secondly, we measured the frequency of searching left and right. The visual search behaviours were classified into three types: mirror searching, visual searching and searching by head turning. Mirror searching means visual search behaviours in which drivers search side-view mirrors. Visual searching means visual search behaviour in which drivers search the side of the car while turning their heads. Searching by turning the head means visual search behaviours in which the driver searches toward the side and rear of a car by turning their heads substantially. From the gyro sensor angle outputs and video recordings, we measured a standard or neutral position of the driver's head and measured the visual search behaviour from there in the right and left directions. Drivers' behaviours at each intersection were measured from the starting point of their deceleration to the centre of the intersection (Figure 19.2).

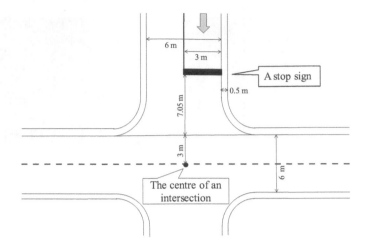

Figure 19.2   Centre of an intersection with stop sign

Driving Route

The driving route included small, non-signalised intersections, with and without stop signs, in a residential area in Kyoto (Japan). We targeted two types of intersections where driving school instructors indicated the necessity to drive carefully in advance. First, two non-signalised intersections with stop signs were selected in order to analyse the driving behaviour while drivers drove straight through the intersection. Additionally, we targeted another intersection without stop signs on the participant's lane, but with stop signs on the intersecting road. The speed limit on this route was 40 km/h.

Procedure

Firstly, participants drove the route once together with an experimenter. Before arriving at the intersections with stop signs, the experimenter told the participant that the next intersection had a stop sign and to consider their safety. Furthermore, flagmen located at road construction sites required participants to stop, in some instances.

Statistical Analysis

Participants' driving speeds, points and visual search behaviours (using the gyro sensor outputs and GPS information), were analysed together. By analysing measures of driving behaviour using gyro sensors (Tada et al., 2008), we calculated the direction, frequency and the maximum angle of search behaviours. In addition, the behavioural measures were complemented by video recordings, as required.

Results

Table 19.1 indicates the points at which drivers started to decelerate or accelerate along with their driving speeds. Table 19.2 indicates the frequency of visual search behaviours.

Table 19.1   Observed points and driving speeds by age group

			with stop sign		without stop sign	
			Elderly	Middle-aged	Elderly	Middle-aged
point (m)	where drivers start to decelerate	M	-41.60	-50.24	-50.27	-47.38
		(SD)	(14.99)	(4.18)	(8.77)	(9.96)
	where drivers start to accelerate	M	0.63	-3.94	-3.92	6.58
		SD	(11.83)	(3.25)	(8.94)	(5.5)
speed (km/h)	where drivers start to decelerate	M	38.33	39.53	40.49	44.78
		(SD)	(4.74)	(4.15)	(4.38)	(4.88)
	30m before the centre of the intersection	M	32.14	31.34	33.56	38.63
		(SD)	(3.66)	(3.7)	(5.97)	(6.75)
	where drivers start to accelerate	M	3.91	0.95	21.50	29.61
		(SD)	(5.10)	(1.48)	(14.42)	(13.54)

Table 19.2   Observed visual search behaviours by age group

direction			with stop sign		without stop sign	
			Elderly	Middle-aged	Elderly	Middle-aged
right	mirror searching	M	0.87	1.13	0.93	0.95
		(SD)	(.83)	(.69)	(.87)	(.69)
	visual searching	M	1.19	1.05	0.85	0.45
		(SD)	(.79)	(.78)	(.86)	(.69)
	searching by turning the head	M	0.39	0.30	0.15	0.00
		(SD)	(.61)	(.44)	(.46)	(.00)
left	mirror searching	M	1.19	0.95	0.67	0.35
		(SD)	(1.02)	(1.06)	(.96)	(.67)
	visual searching	M	0.70	0.63	0.33	0.40
		(SD)	(.56)	(.58)	(.48)	(.68)
	searching by turning the head	M	0.43	1.00	0.11	0.10
		(SD)	(.58)	(1.00)	(.32)	(.31)

Driving Behaviour at Intersections with Stop Signs

The points where drivers started to decelerate differed significantly between age groups (t (31) = 2.85, p < .01). Elderly drivers started to decelerate 42 m before the centre of the intersection, whereas middle-aged drivers decelerated 50 m before the centre of the intersection. However, driving speeds at this point and at the point 30 m before the centre of the intersection did not differ significantly by age group. That is to say, although the driving speeds were the same in both age groups, elderly drivers were later in starting to decelerate and were insufficiently prepared to approach the intersections. Elderly drivers started to accelerate 0.63 m after passing the centre of the intersection, whereas middle-aged drivers started to accelerate 3.94 m before approaching the centre of intersections. However, the age-related differences in points where drivers started to accelerate were not significantly different. Referring to Figure 19.3, 4 m before the centre of an intersection means the point just before reaching the intersection. At this point, elderly drivers were driving at 3.90 km/h, whereas the middle-aged drivers were travelling at 0.95 km/h, which was significantly different (t (45) = 2.51, p < .05). These results indicate that middle-aged drivers completely stopped before reaching an intersection and passed through quickly after reaching the intersection, whereas elderly drives did not stop completely while passing intersections and passed at low speeds. Elderly drivers were also later in reaching the lowest speed and reached the lowest speed at the intersection.

Secondly, a two-way mixed design ANOVA was then conducted to determine the differences in the frequency of searching among age groups and search types. For searching right, the main effect of age group and the interaction term were not significant, but the main effect of searching type was significant (F (2,74) = 13.64, p < .001; Figure 19.3). The lower frequency of searching by turning the head, than the frequency of other searching types, was assessed using the Bonferroni multiple-comparison procedure (p < .001; p < .001). For searching left, the main effect of age group was not significant, but the main effect of searching type and interaction terms were close to being significant (F (2,90) = 2.74, p < .10; F (2,90) = 2.61, p < .10; Figure 19.4. The results of the simple main effect indicated that elderly drivers searched by turning their heads less than middle-aged drivers (p < .05) and that elderly drivers searched more with mirrors than by turning their heads (p < .05). Thus, the frequency of searching by turning the head was inadequate for traffic safety at intersections with stop signs for both age groups. In particular, elderly drivers mainly depended on visual search behaviours without turning their head.

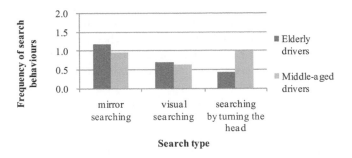

Figure 19.3   Frequency of search right at the intersection with stop sign

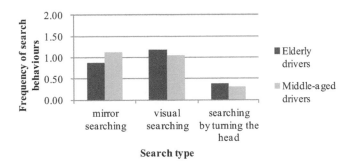

Figure 19.4   Frequency of search left at the intersection with stop sign

Driving Behaviour at Intersections without Stop Signs

For comparison, we examined driving behaviour at intersections without stop signs. We analysed two points, where drivers started to decelerate and where they started to accelerate, except for five elderly drivers whose data were not properly measured. Elderly drivers started to decelerate 50 m before the centre of the intersections, whereas middle-aged drivers did so 47 m before the centre of the intersections. However, the age group difference in the point where drivers started to decelerate was not significant. The driving speeds of elderly drivers at this point and the point 30 m before the centre of an intersection were significantly lower than those of middle-aged drivers (t (45) = 3.17, p < .01; t (45) = 2.73, p < .01). Elderly drivers then started to accelerate 3.92 m after passing the centre of an intersection, whereas middle-aged drivers started to accelerate 6.58 m before reaching the centre of an intersection. The age-related differences in the point where drivers started to accelerate was significant (t (35) = 4.05, p < .001). Before beginning to accelerate, elderly drivers drove at 21.50 km/h, whereas middle-aged drivers drove at 29.61 km/h. The age-related differences in driving speed, before they began accelerating, was significant (t (45) = 1.96, p < .10). These results indicated that elderly drivers drove more slowly than middle-aged drivers while passing intersections. Elderly drivers also reached the lowest speeds before reaching the intersections, whereas middle-aged drivers passed intersections faster and accelerated just after leaving an intersection, which is in contrast to their behaviour at intersections with stop signs.

A two-way mixed design ANOVA was then conducted to determine the differences in the frequency of searching among age groups and by search type. For each direction, the main effects of age group and interaction were not significant, but the main effect of searching type was significant ($F(2,74) = 18.05$, $p < .001$; $F(2,79) = 5.50$, $p < .01$). The results indicated that the frequency of searching by head turning was lower than that of other search types (right $p < .001$; $p < .001$; left $p < .01$; $p < .05$). Thus, the frequency of searching at intersections without stop signs did not differ by age.

Discussion

Driving Behaviour at Intersections with Stop Signs

We conducted field experiments in order to determine the characteristic driving behaviour among elderly drivers at small, unsignalised intersections with stop signs. The results for elderly drivers indicated that they were later in starting to decelerate, although driving speeds before reaching intersections did not differ by age group. Elderly drivers, therefore, need to prepare earlier when approaching intersections. Middle-aged drivers stopped completely before reaching an intersection and passed through more quickly after reaching the intersection, whereas elderly drivers came close to a stop, but did not stop completely at the intersections. Elderly drivers were also later in reaching their lowest speed and came to a stop closer to the centre of the intersections. For intersections where cross roads have the same width, cars on the left should be given preference. Based on this rule, the above behaviour may cause accidents with other cars. The inadequate deceleration could cause more serious workloads to the elderly drivers, because they would have to control the vehicle speed and to conduct a visual search at the same time. It might be a typical type of dual tasks, which can result in increasing risks at intersections with stop signs. Therefore, it is suggested that elderly drivers need to stop completely before reaching intersections and that they need to quickly pass through after reaching the intersections. From the results of the visual search analysis, elderly drivers mainly depended on their mirrors and visual searches that did not involve turning their heads substantially. However, it is vital that drivers engage in visual search behaviours in which the head is turned substantially.

Driving Behaviour at Intersections without Stop Signs

For comparison, we examined driving behaviour at intersections with and without stop signs and driving behaviour at intersections where only one road has a stop sign. Elderly drivers drove at lower speeds than middle-aged drivers, although the points where drivers started to accelerate did not differ between age groups. In contrast to the driving behaviour at intersections with stop signs, elderly drivers started to decelerate before reaching the intersection, whereas middle-aged drivers passed intersections at higher speed and started to accelerate after passing the centre of an intersection even though they had passed at a higher speed. The results for driving speed demonstrate the safe side of elderly drivers and the risky side of middle-aged drivers. It was found that the middle-aged drivers in particular tended to alter their behaviour depending on the presence or absence of a traffic sign. The results of examining visual search behaviours suggested that the frequency of visual search behaviours did not differ by age. This may be because the driving workload when passing straight through an intersection is relatively low. Therefore, it was suggested that age-related differences were small and that the risk for elderly drivers' did not rise substantially at intersections without stop signs.

Conclusions and Future Research

To conclude, we found that driving behaviour differed for intersections with or without stop signs between each age group. This study indicated that elderly drivers had problems controlling speed, did not completely stop at stop signs, and searched without turning their heads substantially. Nevertheless, they normally drove at low speeds and otherwise had relatively safe habits. The results of this study indicate that accidents may be caused in situations where a middle-age driver is driving at high speed without searching at an intersection without a stop sign, while an elderly driver is approaching the same intersection with a stop sign and does not stop completely. Therefore, it is necessary for traffic safety to examine the interactions between multiple cars considering age differences in future research.

Acknowledgment

The authors are grateful to West Nippon Expressway Company, Yamashiro Driving School and Kyotanabe City Silver Human Resources Centre for their help.

References

Ikeda, T., Mori, N., Furuya,H., Minda, H., Ueno, K., Kanto, M., Funakawa, I., Yamanaka, A., and Ichihashi, M. (2004). A case of road traffic factors and measures of accidents caused by elderly drivers. *Proceedings of Infrastructure Planning, 30*, 297. (In Japanese).

Institute for Traffic Accident Research and Data Analysis. (2000). Accidents involving elderly drivers. *ITARDA. Information*, 68.

Oxley, J., Fildes, B., Corben, B., and Langford, J. (2006). Intersection design for older drivers. *Transportation Research Part F: Traffic Psychology and Behaviour, 9*, 335–46.

Okamura, K., and Fujita, G. (1997). Driving performance of older drivers observed in safe driving classes. *Reports of the National Research Institute of Police Science. Research on traffic safety and regulation, 38*(2), 126–35. (In Japanese).

Renge, K., Ishibashi, T., Oiri, M., Ota, H., Tsunenari, S., and Mukai, M. (2003). Elderly drivers' driving performance and hazard perception. *Journal of Japan Applied Psychology, 29*, 62–73. (In Japanese with English abstract).

Tada, M., Okada, M., Noma, H., Iida, K., and Renge, K. (2009). A study of relationships among eye-mark data and gyro data for driving behavior analysis. *The Technical Report of the Image Information and Television Engineers, 33*(54), 33–6. (In Japanese with English abstract).

Tada, M., Segawa, M., Okada, M., Renge, K., and Kogure, K. (2008). Automatic evaluation system of driving skill using wearable sensors and its trial application to safe driving lecture, *The Technical Report of the Institute of Electronics, Information and Communication Engineers, 108*, 1–6. (In Japanese with English abstract).

Chapter 20

Older Drivers' Reasons for Continuing to Drive

Tsuneo Matsuura

Jissen Women's University, Japan

Introduction

Road accident statistics and studies on driving behaviour indicate that driving becomes increasingly risky with age (Catchpole et al., 2005; Cushman, 1996; Evans, 1991; Massie et al., 1995; Ota et al., 2004; Tefft, 2008). Thus, many older drivers modify or self-regulate their driving behaviour to reduce their cognitive overload and to minimize their risk of accidents. The engagement in self-imposed driving restrictions has been attributed to impaired health (Alvarez and Fierro, 2008; Donorfio et al., 2008; Lyman et al., 2001), poor vision (Ball et al., 1998; Lagland et al., 2004; McGwin et al., 2000), poor physical functioning (Lyman et al., 2001; Marottoli et al., 1993), and cognitive impairment (Edwards et al., 2010; Freund and Szinovacz, 2002; Stutts et al., 1998). These medical and functional impairments reduce older people's ability to drive. Research suggests that those who recognize their poor driving ability and who have a lower level of driving confidence are more likely to limit their driving (Baldock et al., 2006; Charlton et al., 2006; Rudman et al., 2006).

The higher accident risk and degraded driving abilities of older drivers may mean that, despite their self-regulation, some older drivers with functional and cognitive difficulties may be more at risk of accident involvement (Christ, 1996; Ross et al., 2009). The proportion of such drivers is highest among the oldest drivers. Self-regulation is beneficial but insufficient for many current older drivers. Although older drivers tend to drive a shorter distance annually than younger drivers, due to self-regulation and changes in lifestyle after retirement, many older drivers still use their vehicles five to six times a week (Ackerman et al., 2008; Catchpole et al., 2005; Lyman et al., 2001; Matsuura, 2010). The aim of the present study was to determine why some older drivers, who should limit or cease driving, continue to drive.

Self-regulation involves four dimensions: driving skill and ability, life and society, self-worth and the role of the automobile as transportation (Donorfio et al., 2009). Declining driving skill and ability due to aging and impaired health may prompt an individual to adopt self-regulatory driving. However, those whose driving keeps them connected to life and society (for example, working and visiting) may resist limiting their driving. Because older drivers tend to view driving as an indicator of independence and self-worth (Donorfio et al., 2009; Rudman et al., 2006), they hesitate to self-regulate their driving. They may also need a car for practical reasons, especially if no public transport is available near their home or no other drivers are available to take them places (Baldock et al., 2006; Catchpole et al., 2005; Lagland et al., 2004; Schlag et al., 1996).

This study used questionnaires and observation to examine factors influencing the older drivers' decision to continue driving actively (meaning almost daily). The purpose was to clarify whether such reasons are acceptable in terms of mobility and safety, and whether less substantial reasons for driving actively increase with age.

Method

Participants

The participants were drivers aged 69 and older ($N = 416$, $M = 74.9$ years) who attended one of seven driving schools to take part in the Older Driver Programme. In Japan, this programme is compulsory for older drivers (70 years and over) in order to renew their driving licence every three years, irrespective of whether they have had any demerit points. The programme includes a driving school instructor's observation of their driving behaviour on a driving course, if their physical condition is not too bad. Participant drivers would be more active than the general population of older drivers of the same age, because all of them attempted to drive on the course.

Studies and Items

This study consisted of three questionnaires and two observations (Table 20.1). Participants answered the questionnaires, as directed by driving school instructors, in a waiting room before the Older Driver Programme began. The first questionnaire examined driver characteristics (for example, gender, age, driving frequency, along with history of accidents and violations). The total accident and violation score was the summation of the five violation and accident variables. Experience of an accident was scored accordingly: a score of 1 = damage-only not responsible accident, 2 = injury not responsible accident, 3 = damage-only responsible accident and 4 = injury responsible accident. The second questionnaire asked the participants to rate their anxiety while driving, by responding to 26 items using a 5-point scale (1 = very seldom, 2 = rather seldom, 3 = sometimes, 4 = often, 5 = very often). The third questionnaire involved participants' self-evaluation of their own driving, using a 4-point scale (1 = False, 2 = False to some extent, 3 = True to some extent and 4 = True). It consisted of 13 items addressing six types of questions on the necessity of driving (3 items, for example, 'I live in an area where

I managed to spend my life without a car'), age-related risky driving (3 items, for example, 'I am afraid I may have missed a traffic-signal'), compensatory driving (2 items, for example, 'I avoid driving an unfamiliar road'), driving confidence (2 items, for example, 'I am confident I can continuing to drive for more than three years'), intention to stop driving (2 items, for example, 'I do not intend to stop driving') and enjoyment of driving (one item, ' I enjoy driving').

Observations were conducted on a driving course at the driving schools where the participants were enrolled in the Older Driver Programme. An instructor observed each participant's driving behaviour while sitting next to the participant in the vehicle. The instructor scored the participant's skill, attention and speed on a 5-point scale (1 = very poor, 2 = poor, 3 = average, 4 = good, 5 = excellent). Participants' posture and understanding ability were also observed in order to evaluate their apparent age.

Table 20.1   Driving characteristics of study subjects for four surveys

Surveys and Characteristics (*n*)		*n* or *M* ± *SD*	%
Questionnaire about driver			
Sex (416)	Male	354	85.1
	Female	62	14.9
Age (416)	69 - 74 years	221	53.1
	75 - 88 years	195	46.9
	Mean ± SD	74.9 ± 4.2	
Driving frequency (416)	Almost every day	265	63.7
	3 or 4 days a week	89	21.4
	1 or 2 days a week	37	8.9
	1 or 2 days a month	10	2.4
	Few days a year	15	3.6
Traffic violations (349)	No violations	293	84.0
	Violations	56	16.0
Injury responsible accidents (350)	No accidents	349	99.7
	Accidents	1	0.3
Injury not responsible accidents (350)	No accidents	349	99.7
	Accidents	1	0.3
Damage-only responsible accidents (350)	No accidents	328	93.7
	Accidents	22	6.3
Damage-only not responsible accidents (350)	No accidents	340	97.1
	Accidents	10	2.9

Table 20.1 *Concluded*

Surveys and Characteristics (*n*)		*n* or *M ± SD*	%
Total accident & violation score (349)	Mean ± SD	0.5 ± 1.1	
Questionnaire of driving anxiety with 5-point scale			
Driving anxiety with 26 items (350)	Mean ± SD	3.0 ± 0.8	
Questionnaire of driving with 4-point scale			
Necessity of driving with 3 items (353)	Mean ± SD	2.3 ± 0.8	
Age-related risky driving with 3 items (346)	Mean ± SD	2.4 ± 0.8	
Compensatory driving with 2 items (351)	Mean ± SD	3.1 ± 0.7	
Driving confidence with 2 items (351)	Mean ± SD	3.2 ± 0.6	
Intention to stop driving with 2 items (350)	Mean ± SD	1.8 ± 0.9	
Enjoyment of driving (351)	Mean ± SD	2.9 ± 0.9	
Observation by instructors (5-point scales)			
In-car observation of driving behaviour with 3 items (412)	Mean ± SD	3.2 ± 0.6	
Apparent age with 2 items (413)	Mean ± SD	2.7 ± 0.7	

Results

Driving Characteristics of the Participants

Table 20.1 lists the driving characteristics of the participants. Most participants drove frequently, and two-thirds drove almost every day, while only 10 per cent of the participants had been involved in an accident in the last three years. The result for the driving anxiety questionnaire, which was answered on a five-point scale indicated that on average participants reported a moderate level of anxiety ($M = 3.0$).

A mean score of 2.3 for the necessity of driving items (4-point scale) indicated that roughly half of the participants did not have a strong need to drive. Participants also reported 'to some extent' conducting compensatory or self-regulatory driving ($M = 3.1$), having confidence in their own driving ($M = 3.2$) and enjoying driving ($M = 2.9$).

Effect of Age on Driving Characteristics

Correlations between age and the driving characteristics are shown in Table 20.2. Six characteristics were weakly but significantly correlated with age. Older drivers exhibited poorer driving behaviour, looked older, self-reported more age-related risky driving, self-reported more compensatory driving, exhibited stronger intentions to stop driving and were more likely to be male. The other six characteristics, including driving frequency and anxiety, did not change significantly with age.

To determine when changes happen, older drivers were divided into four groups: Group 1 (69 to 71 years), Group 2 (72 to 74 years), Group 3 (75 to 79 years) and Group 4 (80 years and older). ANOVA with age group as the independent variable was conducted for each of the six variables (Table 20.2). F tests and multiple comparison tests indicated that the four age groups should be divided into three groups: with Group 1 and Group 4 being considered separately, while groups 2 and 3 were analysed together.

In addition we examined whether older drivers could be divided into a young-old group (Groups 1 and 2) and an old-old group (Groups 3 and 4). Six t-tests were performed to determine whether the two age groups differed from each other on these six variables (Table 20.2). The two groups differed significantly on four variables: observed driving behaviour, apparent age, intention to stop driving and gender. Thus, all following analyses were conducted by dividing participant drivers into a young-old group (69 to 74 years) and an old-old group (75 years and older).

Table 20.2   Relationships between age and other driving characteristics

Variable	Correlation with age	F test among four age groups		t test
			Multiple comparison	Group 1,2 vs 3,4
	r or rs	F	Group 1 vs Group 4	Group 2 vs Group 3	t
Sex	-.18*	5.00**	**	n.s	3.36**
Age-related risky driving	.17**	3.49*	*	n.s	-1.45
Compensatory driving	.15**	2.57	*	n.s	-1.45
Intention to stop driving	.27**	6.71**	**	n.s	-3.78**
Observed driving behaviour	-.11*	2.33		n.s	2.31*
Apparent age	.17**	4.62**	*	n.s	-3.18**

Note: Four age groups comprise Group 1 (69–71years), Group 2 (72–74 years), Group 3 (75–79 years) and Group 4 (80+ years). r_s denotes Spearman rank order correlation.

* $p < .05$. **$p < .01$.

Comparison of the Reasons for Driving between Active and Less Active Drivers among the Young-old and the Old-old

The reasons for driving were compared between a group of active young-old drivers (Group A, n = 148) and a group of less active young-old drivers (Group B, n = 71), using logistic regression (Table 20.3). The dependent variable was group membership (whether they were in group A or B). Ten predictor variables were used: overall accident and violation score, anxiety, necessity of driving, age-related risky driving, self-reported compensatory driving, driving confidence, intention to stop driving, enjoyment of driving, observed driving behaviour and apparent age. Among these 10 predictor variables, three variables differed significantly between groups A and B. Group A (active young-old drivers) felt less anxious while driving, were observed to drive more safely, and needed a car much more than Group B (less active young-old drivers). Group A drivers drove more actively than Group B drivers, probably because they had more substantial reasons for driving.

The same analysis was used to compare reasons for driving between the active old-old drivers (Group C, n = 116) and the less active old-old drivers (Group D, n = 79), but none of the predictor variables significantly differentiated between these two groups. The results indicate that Group C drivers drive actively and Group D drivers drive less actively for the same reasons. Group C drivers may continue to drive actively without sufficient reason, or Group D drivers may restrict their driving despite not having sufficient reason.

Table 20.3   Variables explaining differences between Group A and B

Explanatory variable	B	SE	Wald test		Exp (B)
Total acci. and violation score	0.01	0.19	0.00		1.00
Anxiety	0.02	0.01	3.93	*	1.02
Necessity of driving	-0.31	0.10	8.87	**	0.73
Age-related risky driving	-0.16	0.10	2.26		0.85
Compensatory driving	0.09	0.14	0.38		1.09
Driving confidence	-0.16	0.21	0.57		0.85
Intention to stop driving	0.20	0.15	1.71		1.22
Enjoyment of driving	0.04	0.24	0.03		1.04
Observed driving behaviour	-0.45	0.17	7.11	**	0.64
Estimated apparent age	-0.29	0.18	2.70		0.75
Constant	5.66	3.00	3.56		287.79

* $p < .05$. ** $p < .01$.

To determine whether Group C drivers were more like Group A or Group B drivers, a one-way ANOVA was performed with age group (Group A, Group B and Group C) as the independent variable and with driving anxiety, observed driving behaviour and necessity of driving as dependent variables. ANOVA with driving anxiety as a dependent variable did not indicate significant differences among the three groups ($F(2, 293) = 1.34$, *ns*).

However, observed driving behaviour differed significantly between the groups ($F(2, 331) = 5.98$, $p < .01$) with post hoc comparisons (Games-Howell test) indicating significant differences existed between groups A and C, as well as between groups A and B, but not between groups B and C. These results suggest that the driving behaviour of Group C drivers was similar to that of Group B drivers. Figure 20.1 presents the mean observed driving behaviour scores for the four groups, including Group D.

For necessity of driving, the ANOVA indicated significant differences among the three groups ($F(2, 286) = 6.33$, $p < .01$). Post hoc tests (Tukey HSD) indicated significant differences between groups B and C, as well as between groups A and B (Figure 20.2). However, groups A and C did not differ significantly.

In summary, Group C drivers, like Group A drivers, seemed to drive actively because driving was a necessity for them. However, compared with Group A drivers the driving ability of Group C and Group B drivers was poor.

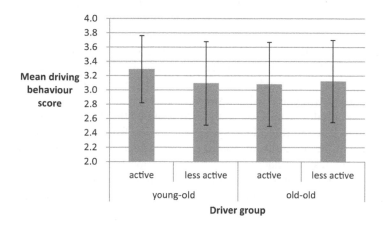

Figure 20.1   Mean observed driving behaviour score for each driver group

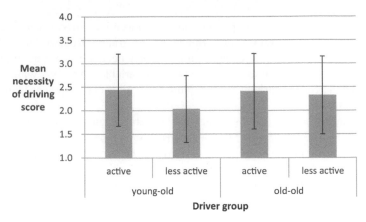

Figure 20.2   Mean necessity of driving score for each driver group

Classification and Characteristics of the Four Groups

Observed driving behaviour and the necessity of driving were key variables in determining the differences among groups A, B, C and D. Thus, we classified each of these four groups, which consisted of two categories of age and two categories of driving frequency, by observed driving behaviour (bad, good) and by the necessity of driving (low, unknown, high) (Table 20.4). An observed driving behaviour score of less than or equal to three (in other words, the scale mid-point) was classified as poor driving behaviour, while a score of more than three was classified as good driving behaviour. For the necessity of driving scale, a score of less than or equal to 2.5 (mid-point) was classified as low necessity, while a score of more than 2.5 was classified as a strong necessity.

SPSS was used for log-linear analysis to examine the relationships among the four variables listed in Table 20.4 (age group, driving frequency, observed driving behaviour and the necessity of driving). Log-linear analysis is an extension of the chi-square test of independence and is a powerful tool for determining relationships among variables in multi-way contingency tables. The purpose of log-linear analysis is to find the most parsimonious model that can account for cell frequencies in a table. The term 'log-linear' derives from the fact that the dependent variable is the logarithm of the cell frequencies explained by the main effects and interactions of the variables.

A hierarchical log-linear analysis with the backward elimination strategy revealed three significant interactions: age group × driving frequency × observed driving behaviour, driving frequency × observed driving behaviour × necessity of driving and age group × necessity of driving.

The three-way interaction among age group, driving frequency and observed driving behaviour indicates that active drivers showed better driving behaviour than less active drivers in the young-old group (97/149 (65 per cent) versus 35/71

Table 20.4   Classification of each group by driving behaviour and necessity of driving

Group	Bad driving behaviour			Good driving behaviour		
	Necessity			Necessity		
	Low	Unknown	High	Low	Unknown	High
Active young-old (A: *n* = 149)						
n	28	2	22	51	11	35
%	18.8	1.3	14.8	34.2	7.4	23.5
Less active young-old (B: *n* = 71)						
n	21	7	8	22	4	9
%	29.6	9.9	11.3	31.0	5.6	12.7
Active old-old (C: *n* = 113)						
n	25	10	24	25	12	17
%	22.1	8.8	21.2	22.1	10.6	15.0
Less active old-old (D: *n* = 78)						
n	15	13	8	23	2	17
%	19.2	16.7	10.3	29.5	2.6	21.8

(49 per cent), χ^2 (1) = 5.00, $p < .05$), but not in the old-old group (54/113 (48 per cent) versus 42/78 (54 per cent), χ^2 (1) = 0.68, *n.s*). The three-way interaction among driving frequency, observed driving behaviour and necessity of driving indicates that active drivers needed a car for driving more than less-active drivers in the bad driving behaviour group (46/111 (41 per cent) versus 16/72 (22 per cent), χ^2 (1) = 12.00, $p < .01$), but not in the good driving behaviour group (52/151 (34 per cent) versus 26/77 (34 per cent), χ^2 (1) = 2.86, *n.s*). The two-way interaction between age group and necessity of driving was also significant (χ^2 (2) = 6.72, $p < .05$). A higher proportion of the young-old drivers did not need a car as much for driving as did the old-old drivers (122/220 – 55 per cent versus 88/191 – 46 per cent).

Reasons for Active Driving in Spite of Bad Driving and Low Necessity of Driving

Some drivers in Group A (*n* = 28, 19 per cent) and Group C (*n* = 25, 22 per cent) drove almost every day, even though they were observed to be bad drivers and did not need to drive that much (See Table 20.4). In order to clarify their reasons for active driving, one-way ANOVAs were conducted with four categories of drivers (meaning bad drivers with low necessity, bad drivers with high necessity, good drivers with low necessity and good drivers with high necessity) as the independent variable. Gender, total accident and violation scores, compensatory driving, driving confidence, intention to stop driving, enjoyment of driving and apparent age were the dependent variables.

ANOVAs for Group A indicated that only compensatory driving and apparent age were significantly different among the four categories of drivers (F (3, 134) = 3.75, $p < .05$; F (3, 135) = 10.29, $p < .01$). Multiple comparison tests (Tukey's HSD) to determine specific differences among the four categories of drivers indicated that bad drivers with a low necessity to drive were most likely to exhibit compensatory driving ($M = 3.25$) and looked the oldest ($M = 3.09$), while good drivers with a high driving-necessity were the least likely to exhibit compensatory driving ($M = 2.61$) and looked the youngest ($M = 2.30$).

ANOVAs for Group C indicated that only enjoyment of driving and apparent age were significantly different among the four categories of drivers (F (3, 90) = 3.02, $p < .05$; F (3, 90) = 4.56, $p < .01$). Multiple comparison tests (Tukey's HSD) revealed that bad drivers were less likely to enjoy driving ($M = 2.80$) and looked older ($M = 3.02$), whereas good drivers were more likely to enjoy driving ($M = 3.31$) and looked younger ($M = 2.53$).

Discussion

Two-thirds of the participants ($n = 265$, 64 per cent) drove almost every day and were defined as active drivers. The other participants, who were current drivers but drove less than five days per week, were defined as less active drivers. When we divided the participants into young-old (69 to 74 years) and old-old (75 years and older), the percentage of active drivers was higher in the young-old group (68 per cent) than in the old-old group (59 per cent). The results indicating that many older drivers, especially young-old drivers, drove almost every day were consistent with the findings of previous research (Ackerman et al., 2008; Catchpole et al., 2005; Lyman et al., 2001).

The main purpose of this study was to examine whether active drivers, especially active old-old drivers, drive for acceptable reasons. A comparison of the reasons for driving between the active and less active drivers indicated that active drivers in the young-old group reported less anxiety while driving, drove better, and needed a car much more than less active drivers. Active and less active old-old drivers did not differ in their reasons for driving. Log-linear analysis replicated the finding that among the young-old group, active drivers demonstrated better driving behaviour than less active drivers. However, this trend was not observed for old-old drivers. These results imply that active young-old drivers have some substantial reasons to drive, whereas old-old drivers do not.

Of course, some drivers in Group C (active old-old drivers) had good driving ability ($n = 54$, 48 per cent), as was the case for the less active old-old drivers ($n = 42$, 54 per cent). The problem group was the remaining drivers in Group C who drove actively in spite of their poor driving ability ($n = 59$, 52 per cent). This group did not appear to realize that they are high-risk drivers, because their compensatory driving and driving confidence did not differ significantly from the other drivers in Group C. They were also less likely to enjoy driving and looked older than the good drivers in Group C. Although a small proportion of them ($n = 24$, 21 per cent) had a high necessity of driving score, the remainder did not.

Some drivers may continue to drive actively for reasons other than those considered in this study (Donorfio et al., 2009; Rudman et al., 2006). They may continue to drive because of a long-standing identity with operating a car. They may continue to drive by force of habit, as they have not been involved in road accidents.

Finally, some measures should be taken for active old drivers with unacceptable reasons for driving. Providing them with feedback on their driving ability and driving style (for example, after a driving test, driving lesson or after a self-assessment test) may help them to realize that they should no longer be active drivers (Eby et al., 2003; Matsuura, 2010; Matsuura et al., 2008). Screening tests for cognitive impairment, which have been part of the licence renewal process in Japan since 2009, may also be useful in preventing them from actively driving.

References

Ackerman, M.L., Edwards, J.D., Ross, L.A., Ball, K.K., and Lunsman, M. (2008). Examination of cognitive and instrumental functional performance as indicators for driving cessation risk across 3 years. *The Gerontologist, 48*, 802–10.

Alvarez, F.J., and Fierro, I. (2008). Older driver, medical condition, medical impairment and crash risk. *Accident Analysis and Prevention, 40*, 55–60.

Baldock, M.R.J., Mathias, J.L., McLean, A.J., and Berndt, A. (2006). Self-regulation of driving and its relationship to driving ability among older adults. *Accident Analysis and Prevention, 38*, 1038–45.

Ball, K., Owsley, C., Stalvey, B., Roenker, D.L., Sloane, M.E., and Graves, M. (1998). Driving avoidance and functional impairment in older drivers. *Accident Analysis and Prevention, 30*, 313–22.

Catchpole, J., Styles, T., Pyta, V., and Imberger, K. (2005). *Exposure and accident risk among older drivers*. Research Report ARR 366. Vermont South, Australia: ARRB Group Ltd.

Charlton, J.L., Oxley, J., Fildes, B., Oxley, P., Newstead, S., Koppel, S., and O'Hare, M. (2006). Characteristics of older drivers who adopt self-regulatory driving behaviours. *Transportation Research Part F, 9*, 363–73.

Christ, R. (1996). Ageing and driving: Decreasing mental and physical abilities and increasing compensatory abilities? *IATSS* Research, 20, 43–52.

Cushman, L.A. (1996). Cognitive capacity and concurrent driving performance in older drivers. *IATSS* Research, 20, 1, 38–45.

Donorfio, L.K.M., D'Ambrosio, L.A., Coughlin, J.F., and Mohyde, M. (2008). Health, safety, self-regulation and the older driver: It's not just a matter of age. *Journal of Safety Research, 39*, 555–61.

Donorfio, L.K.M., D'Ambrosio, L.A., Coughlin, J.F., and Mohyde, M. (2009). To drive or not to drive, that isn't the question: The meaning of self-regulation among older drivers. *Journal of Safety Research, 40*, 221–6.

Eby, D.W., Molnar, L.J., Shope., J.T., Vivoda, J.M., and Fordyce, T.A. (2003). Improving older driver knowledge and self-awareness through self-assessment: The driving decisions workbook. *Journal of Safety Research, 34*, 371–81.

Edwards, J.D., Bart, E., O'Connor, M.L., and Cissell, G. (2010). Ten years down the road: Predictors of driving cessation. *The Gerontologist, 50*, 3, 393–9.

Evans, L. (1991). *Traffic Safety and the Driver*. New York, USA: Van Nostrand Reinhold.

Freund, B., and Szinovacz, M. (2002). Effects of cognition on driving involvement among the oldest old: Variations by gender and alternative transportation opportunities. *The Gerontologist, 42*, 621–33.

Lagland, D.R., Satariano, W.A., and MacLeod, K.E. (2004). Reasons given by older people for limitation or avoidance of driving. *The Gerontologist, 44*, 237–44.

Lyman, J.M., McGwin, G., and Sims, R.V. (2001). Factors related to driving difficulty and habits in older drivers. *Accident Analysis and Prevention, 33*, 413–21.

Marottoli, R.A., Ostfeld, A.M., Merrill, S.S., Perlman, G.D., Foley, D.J., and Cooney, Jr. L.M. (1993). Driving cessation and changes in mileage driven among elderly individuals. *Journal of Gerontology: Social Sciences, 48*, S255–S260.

Massie, D.L., Campbell, K.L., and Williams, A.F. (1995). Traffic Accident involvement rates by driver age and gender. *Accident Analysis and Prevention, 27*, 73–87.

Matsuura, T. (2010). Older drivers' risky and compensatory driving: Development of a safe driving workbook for older drivers. In D. Hennessy (Ed.), *Traffic Psychology: An International Perspective*. New York, USA: Nova Science Publishers.

Mastsuura, T., Ishida, T., and Mori, N. (2008). Safe driving workbook for older drivers (In Japanese). Tokyo, Japan: Kigyou-Kaihatsu Center.

McGwin, G. Jr., Chapman, V., and Owsley, C. (2000). Visual risk factors for driving difficulty among older drivers. *Accident Analysis and Prevention, 32*, 735–44.

Ota, H., Ishibashi, T., Oiri, M., Mukai, M., and Renge, K. (2004). Self-evaluation skill among older drivers (In Japanese). *Japanese Journal of Applied Psychology, 30*, 1–9.

Ross, L.A., Clay, O.J., Edwards, J.D., Ball, K.K., Wadley, V.G., Vance, D.E., and Joyce, J.J. (2009). Do older drivers at-risk for crashes modify their driving over time? *Journal of Gerontology: Psychological Sciences, 64B*, 163–70.

Rudman, D.L., Friedland, J., Chipman, C., and Sciotino, P. (2006). Holding on and Letting go: The perspectives of pre-seniors and seniors on driving self-regulation in later life. *Canadian Journal of Ageing, 25*, 65–76.

Schlag, B., Schwenkhagen, U., and Trankle, U. (1996). Transportation for the elderly: Towards a user-friendly combination of private and public transport. *IATSS* Research, *20*, 75–82.

Stutts, J.C., Stewart, J.R., and Martell, C. (1998). Cognitive test performance and crash risk in an older driver population. *Accident Analysis and Prevention, 30*, 337–46.

Tefft, B.C. (2008). Risks older drivers pose to themselves and to other road users. *Journal of Safety Research, 39*, 577–82.

PART V
Driver Behaviour and Driving Simulation

Chapter 21

A Tandem Model of Proceduralization (Automaticity) in Driving[1]

Samuel G. Charlton and Nicola J. Starkey

Traffic and Road Safety Research Group, University of Waikato, New Zealand

Introduction

This chapter describes our thoughts on the nature of everyday driving, with a particular emphasis on the processes that govern driver behaviour in familiar, well-practised situations. Based on consideration of established models of driver behaviour and an initial study of proceduralized driving in a driving simulator we describe how drivers' vehicular control and attention to road features are modified with repeated exposure. We have called this approach a 'tandem model' as it includes both explicit and implicit processes involved in driving performance.

The Nature of Driving

Driving is one of the most widely practised skills in the adult population, and as such, its examination can provide us with insights about the acquisition and performance of skilled behaviour more generally. Many models of driver behaviour have been proposed over the years. Some of them have focused on the manual control aspects, describing driving in terms of servo-control theory (McRuer and Weir, 1969; Reid, 1983; Sheridan, 2004). Other models have approached driving from an information-processing perspective, focusing on the role of visual search, attention, and mental load in determining how drivers obtain and process information about potential hazards (Brown, 1962; Hancock et al., 1990; Lee and Boyle, 2007; Recarte and Nunes, 2003; Theeuwes, 2000). Another group of models have used the idea of motivational factors that involve feedback loops or thresholds prompting drivers to adjust their behaviour on the basis of their perceptions of risk or task difficulty (Fuller, 2005; Lewis Evans and Rothengatter, 2009; Näätänen and Summala, 1974; Taylor, 1964; Wilde, 1982, 2002). Finally, other models have described the driving task as being comprised

---

1 Paper presented at the Symposium on the Contemporary Use of Simulation in Road Safety, International Congress of Applied Psychology, 15 July 2010. Originally titled *Driving Without Awareness: Insights into the Unconscious Driver*.

of several hierarchical levels of complexity ranging from simple control skills to rule-based and goal-directed behaviour associated with the driving task (Michon, 1989; Ranney 1994).

While these models have made valuable contributions in guiding research and engineering practices, few of them have addressed the effects of repeated exposure to a particular road or traffic situation on driving performance. The absence of this component is important because extended practice would appear to produce some qualitative changes in skilled performance. From skills as diverse as acting to equitation, it has been noted the effect of prolonged practice is to transform effortful and deliberate approximation into effortless proficiency (Fitts and Posner, 1967; Groeger, 2002; Morris, 1981). Indeed, driving familiar routes has often been identified as an example of behaviour that becomes proceduralized or automated with extended practice (Fitts and Posner, 1967; MacKay, 1982; Norman, 1981; Shiffrin and Schneider, 1977), and yet relatively little is known about the process of proceduralization in driving or how the resulting driving performance may differ from driving with what William James (1890) called 'active attention'.

Automatic performance has generally been characterized as efficient, unintentional, unconscious behaviour that has become proceduralized through extensive practice (Moors and De Houwer, 2006; Shiffrin and Schneider, 1977). Once proceduralized, these automatic behaviours are often initiated and executed in a stimulus-driven or 'bottom-up' fashion, with little or no demand on conscious attentional resources (Logan, 1988). Automatic processes have also been defined by distinguishing them from explicitly controlled, attentionally demanding, effortful, knowledge-based performance, often referred to as 'top-down processing' (Theeuwes, 2010). It has been suggested that both types of cognitive processing, top-down (conscious) and bottom-up (automatic), can be seen in driver behaviour (Charlton, 2004; Ranney, 1994; Summala, 1988; Trick et al., 2004). The top-down mode, involving conscious deliberation and effort, is experienced by learner drivers or even experienced drivers in an unfamiliar environment or while concentrating on a demanding manoeuvre such as overtaking. The 'bottom-up' mode, composed of well-rehearsed perception-action units, enables experienced drivers to maintain their speed, lane position, following distances, and negotiate traffic with little or no conscious attention or effort. As early as 1938 Gibson and Crooks noted that: 'The relative importance of effortful attention and of habit in safe driving needs to be worked out' (p. 458). Unfortunately, this is largely still the case.

The Problem of Demonstrating Automaticity

Driving would appear to provide an excellent domain for the investigation of cognitive automaticity, both because it is a well-practised skill in the general population and because it is composed of a diverse range of attention, perception, decision-making, and motor control components. In spite of its apparent suitability, however, the preponderance of automaticity research has employed more constrained laboratory tasks such as repetition priming, go/no-go reaction time, subitizing, and text processing (Logan, 1990; Naparstek and Henik, 2010; Rawson and Middleton, 2009; Verbruggen and Logan, 2009). Similarly, few studies of driver behaviour have attempted to investigate proceduralized driving. Most experimental studies of driver behaviour have relied on short, hour-long experimental sessions in which participants must adapt to novel driving experiences in unfamiliar vehicles or driving simulators with the knowledge that their performance is being observed and analysed, circumstances that might inhibit drivers' inclination to disengage their active attention and rely on habitual or automatic driving performance.

It should be noted that recent studies of naturalistic driving have used vehicle sensors and unobtrusive video cameras to record drivers' behaviour over periods ranging from one week to one year (Dingus et al., 2006; Stutts et al., 2005). These studies, however, have focused on the identification of precursors to crashes and near crashes (Dingus et al., 2006) or the occurrence of driver distractions (Stutts et al., 2005) and in these cases the proceduralized form of 'driving without awareness' is often treated as an exception to 'normal' driving; an undesirable state that occurs when a driver fails to pay attention. More typically, however, the investigation of expertise in driving has involved cross-sectional experimental designs comparing the performance of expert and novice drivers on simulated or actual driving tasks (Borowsky et al., 2008; Duncan et al., 1991; Galpin et al., 2009; Shinar et al., 1998). Although a range of differences between novice and expert drivers have been noted in these studies, the experimental sessions have again often been short with artificial or unfamiliar situations. Based on these cross-sectional studies some researchers have questioned whether any component of the driving task can be considered truly automatic in the strict sense of the term; in other words effortless, unintentional, and impervious to secondary task disruption (Groeger and Clegg, 1997).

The Effects of Practice and Proceduralization

In one recent longitudinal study, Martens and Fox (2007) investigated the effect of practice on top-down scanning and attention in a desk-top simulator over a period of five days. The participants in this study were reported as increasing their speeds and decreasing their glance duration as the study progressed, with the largest changes taking place on the first day of practice. Interestingly, more practice led to better recollection of the traffic signs along the route but poorer driving performance and post-drive recognition accuracy when a target sign was changed from a priority crossing to a yield sign. The researchers concluded that the effects of practice were to establish 'top-down control' over visual scanning patterns that resulted in a failure to detect changes to the road environment. The researchers also posed the intriguing question of what type of stimulus would prompt drivers to switch back to an active information processing state and proposed that future research should investigate this aspect.

The results of the Martens and Fox (2007) study suggest that extended practice with a particular driving situation, in addition to experience with the driving task more generally, may be necessary to differentiate top-down and bottom-up aspects of driving behaviour. In order to explore this possibility further we conducted a pilot experiment in which participants repeatedly drove a single road in a driving simulator over three months. The principal research questions were: (1) can proceduralized (automatic) driving be produced and detected in a driving simulator with extended practice; and if so, (2) how is automaticity reflected in drivers' behaviour?

Two groups of participants were recruited for this pilot experiment; the first group of nine participants, four males and five females, were assigned to the 'Expert' group who were recruited to drive 20 sessions in a driving simulator. A second group of 12 participants, five males and seven females, formed a 'Casual' group recruited to participate for a single experimental session. The drivers in both groups held unrestricted driving licences and had an average of 11.52 years of driving experience (range 2–35 years, SD = 9.99). The participants drove a 24 km-long simulated road composed of straights and gentle horizontal and vertical curves with realistic road geometry, road markings, and traffic and contained many landmarks (for example, houses, shops, farms, a bridge, and a tunnel) to facilitate participants' recognition of their surroundings over repeated sessions (see Figure 21.1). The driving simulator

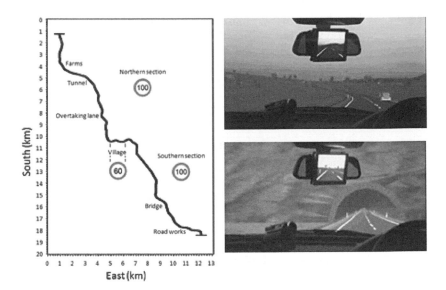

Figure 21.1   An overview of the experimental protocol including: a map of the simulated road (left); two views from the driver's position showing an example of the Volkswagen beetle target (top right) and the tunnel (bottom right)

consisted of a complete automobile positioned in front of angled projection surfaces (as described in Charlton, 2009).

Participants in the Expert group drove the simulated road during 20 experimental sessions, with each session consisting of two 12-km scenarios separated by a short break (see Table 21.1). In each scenario half the road was driven in either of two directions (starting at either the north end or the southern end) but from session to session the simulated road differed only in the placement of other traffic on the road. On some occasions, however, participants encountered sessions containing a Road Works scenario, Change Detection scenario (changes to the landmarks), or Unfamiliar Road scenario (changes in

Table 21.1   The order of the experimental scenarios for each group

	First scenario	Second scenario
Session	Expert group	
1	Northern section, southbound	Southern section, southbound
2	Southern section, northbound	Northern section, northbound
3	Northern section, northbound	Northern section, southbound
4	Southern section, northbound	Southern section, southbound
5	Roadworks - southern section, southbound	Roadworks - southern section, northbound
6	Northern section, southbound	Roadworks - southern section, southbound
7	Northern section, southbound	Change detection - northern section, northbound
8	Unfamiliar road - northern and southern sections, southbound	
9	Northern section, northbound	Northern section, southbound
10	Southern section, northbound	Northern section, northbound
11	Southern section, northbound	Southern section, southbound
12	Northern section, southbound	Northern section, northbound
13	Roadworks - southern section, northbound	Roadworks - southern section, southbound
14	Northern section, southbound	Conversation - southern section, southbound
15	Southern section, northbound	Change detection - northern section, northbound
16	Unfamiliar road - northern and southern sections, southbound	
17	Southern section, northbound	Southern section, southbound
18	Northern section, southbound	Southern section, southbound
19	Northern section, northbound	Conversation - northern section, southbound
20	Southern section, northbound	Northern section, northbound
	Casual group	
1	Roadworks - southern section, northbound (same as session 13 above)	Change detection - northern section, northbound (same as session 15 above)

visual landscape and removal of all familiar landmarks although with identical road geometry). During each drive the participants were asked to drive as they normally would in their own cars and to flash their headlights and say aloud whatever they noticed anything that was interesting, unusual, hazardous, or a Volkswagen beetle (two were included in each session).

Effects on Vehicle Control

As the sessions progressed, the participants developed increasingly stereotyped driving patterns. One indication of this can be seen in the participants' speed and lane position variability. The top-left panel of Figure 21.2 shows the Expert group's speed variability (average absolute deviation in km/h) for a 100 m straight section of road located midway between the start and end of the southern road section, plotted for sessions containing southern road scenarios. Most of the participants' speed variability decreased rapidly in the early sessions and remained low, with the interesting exception of the Unfamiliar road and Conversation scenarios. In the Unfamiliar road session the visual appearance of the road was changed (although the road geometry was identical) and during the Conversation scenario the participants were engaged in a hands-free cell phone conversation as they drove. One of the participants had a distinctly different reaction to these scenarios; their speed variability generally increased over the sessions, and was lowest during the Unfamiliar road and Conversation sessions. Removing that participant's data from the group resulted in the lower-left panel shown in Figure 21.2; a distinct and rapid decrease in mean speed variability, and a return to first session levels during the Unfamiliar road and Conversation scenarios.

As shown in the top-right panel of Figure 21.2, a similar pattern was seen in the participants' lane position variability (average absolute deviation in metres), a tendency for the lane positions to become less variable across sessions, with peaks of high variability during the Unfamiliar road and Conversation scenarios. Finally, in the lower-right panel of Figure 21.2 the speed and lane position variability for the participants in the Casual group (during the first scenario of their session) is compared to that of the Experts during the same scenario (first scenario of Session 13). Participants in the Expert group displayed lower speed and lane position variability than did participants in the Casual group. Statistical analysis with a one-way multivariate analysis of variance indicated a significant difference between the two groups across the two measures [Wilks' Lambda = .666, $F_{(2, 17)}$ = 4.270, $p < 0.05$, $\acute{\eta}_p^2$ (Partial eta squared) = .334], with univariate tests indicating this was largely due to the speed variability measure [$F_{(1, 18)}$ = 5.251, $p < 0.05$, $\acute{\eta}_p^2$ = .226] rather than the lane position variability measure [$F_{(1, 18)}$ = 1.947, $p > 0.10$, $\acute{\eta}_p^2$ = .098].

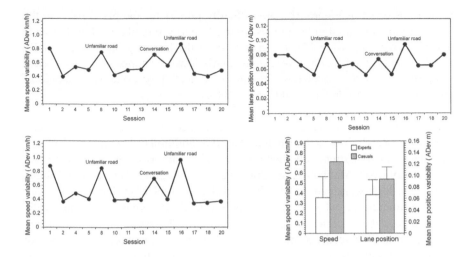

Figure 21.2   Speed variability (across a 100 m section of the southern half of the simulated road) for all participants in the Expert group is shown in the top left panel, and again with one participant removed in the lower left panel. Lane position variability across sessions for all participants in the Expert Group (top right panel). The lower left panel shows mean speed variability and lane position variability for the first scenario driven by the Casual group compared to the Expert group on the equivalent scenario (the first scenario of session 13). Error bars show 95 per cent confidence intervals

Effects on Driving Difficulty

These changes in speed and lane position variability were mirrored in the participants' ratings of driving difficulty and their verbal reports. As can be seen in the left panel of Figure 21.3, the participants in the Expert group rated the driving difficulty of the 20 sessions progressively lower as the experiment progressed. The reductions in variability also coincided with verbal reports that the participants were devoting less attention to the driving task. Comments such as '*I found myself going into auto, not paying attention to what I was doing*' (Participant 2, Session 5), '*Feels very normal, just like the drive home; thinking mostly about food*' (Participant 3, Session 6) and '*Was daydreaming, a lot on my mind*' (Participant 8, Session 7) suggests to us that some withdrawal of attention (or proceduralization) was occurring with repeated practice. Once again, the exceptions to this were the two sessions in which the visual appearance of the road was changed (Unfamiliar road) and two sessions in which they engaged in a hands-free cell phone conversation as they drove (Conversation scenarios). Shown in the right panel of Figure 21.3 are difficulty ratings for the participants in the Casual group compared to the ratings for the Expert group obtained for the same driving scenario (Session 15). The ratings from the Expert group (mean = 1.19, where 1 = easy; no difficulty) were significantly lower than those from the Casual group (mean = 3.00, moderately difficult) as indicated by a one-way analysis of variance; $F_{(1, 18)} = 18.962$, $p < 0.001$, $\acute{\eta}_p^2 = .513$.

It is perhaps not surprising to find that the introduction of cell phone conversation to the driving task resulted in increased ratings of driving difficulty; previous research has demonstrated that the additional attentional demands resulting from concurrent cell phone conversations are associated with increases in driver workload (Matthews et al., 2003; Patten et al., 2004; Törnros and Bolling, 2006). The high workload ratings obtained during the Unfamiliar road sessions are more interesting; why should driving a visually different but geometrically identical road be experienced as more difficult? The participants volunteered that the Unfamiliar road was more difficult because it contained more curves or traffic (it did not). These results suggest that the changes to the appearance of the road alone somehow nullified the effects of extended practice and reinstituted the variability in vehicle control and feelings of driving difficulty encountered in the early sessions of the experiment.

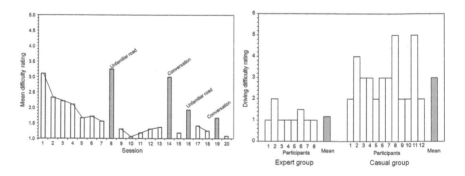

21.3   **The left panel shows the mean ratings of driving difficulty (1 = no difficulty; 7 = extremely difficult) by participants in the Expert group for each session. The right panel shows the Expert group participants' ratings during session 15 compared to the Casual group's ratings of the same scenario in their first session**

Effects on Visual Detection

The participants' performance on the embedded detection task (flashing their headlights and reporting what they noticed while they drove) also became increasingly stereotyped. With practice, participants reduced the total number of things they reported but did report their every encounter with a few specific items (for example, road bumps, an intersection, the tunnel, or large trucks), the same items during every scenario. The left panel of Figure 21.4 shows the decline in the number of new items reported (items not reported in the immediately preceding session) in each session, with the noteworthy exception of scenarios that were physically changed for the Road works, Unfamiliar road, and Change Detection sessions. The right panel of Figure 21.4 shows the average number of all items reported (excluding Volkswagens) by the participants in the Casual group as compared to the participants in the Expert group on the corresponding scenarios (the first scenario of Session 13 and second scenario of Session 15). The pattern of results indicated that overall more items were reported by the Casual participants, but during the Road works scenario (presented first) the participants in the Expert group reported many fewer items than the Casual participants, but during the Detection scenario (presented second) they reported more than the Casual participants (meaning, a significant group X scenario interaction; $F_{(1, 18)} = 11.210$, $p < 0.01$, $\acute{\eta}_p^2 = .384$). It is worth noting that although many changes to the landscape in the Detection scenarios were reported by most of the Expert group, other items were apparently not as noticeable; changes to signage (including changing the wording from English to German) were not reported by any of the participants. This failure to detect changes to road signs is similar to the finding reported by Martens and Fox (2007) where participants did not detect changes to intersection priority signs or may simply reinforce previous findings that participants have a general attentional neglect for familiar road signs (Charlton, 2006).

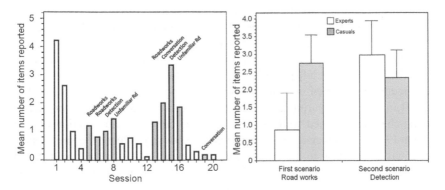

Figure 21.4 **The panel at left shows the mean number of new items (not reported in the preceding session) reported by participants in the Expert group in each session. The right panel shows the mean number of items reported (excluding Volkswagen targets) by the Casual group compared to the number of items reported by the Expert group during equivalent scenarios (the first scenario of session 13 and second scenario of session 15). Error bars show 95 per cent confidence intervals**

The distances at which the Expert participants detected the Volkswagen beetles generally increased across the 20 sessions; the mean distances increased from a low of 30.30 m in Session 2 to a high of 84.59 m in Session 18. There were, however, subtle differences in detection between the first and second scenario of each session. The distances steadily increased for the first scenario presented in each session (sometimes a northern scenario, sometimes a southern scenario), detection during the second scenario, however, decreased initially then increased and exceeded the first scenario distances towards the latter sessions (68.92 m for the first scenario of session 18 compared to 100.28 m for the second scenario). One possible explanation for the first and second scenario differences came from the participants themselves who commented that during the second scenario of each session they were more likely to 'let their mind wander'. Thus in the early sessions Volkswagen detection was best during the first scenario, when the participants were more likely to pay closer attention to the task. With extended practice, however, Volkswagen detection may have become so habitual that little effort was required. Indeed, more than one participant commented that they often flashed their headlights (or found themselves reaching for them) when they saw Volkswagen beetles while driving their own cars. The long detection distances eventually present in the second scenario suggest that detection in this automatic mode might be faster than effortful top-down detection that was more likely to occur in the first scenarios of the early sessions.

Effects on Speed Regulation

Another aspect of the Expert group's driving performance changed with extended practice was an increasing tendency to reduce their speed as they entered a 400 m tunnel included in the northern section of the simulated road. Figure 21.5 shows the participants' speeds during their southbound northern road scenarios. As can be seen in the left panel of the figure, the participants displayed larger speed reductions as their experience with the tunnel increased [$F_{(1, 7)}$ = 33.198, p < 0.001, $\acute{\eta}_p^2$ = .826]. During the Unfamiliar road scenario no tunnel was present and no changes in speed were observed for the corresponding section of road. Session 19 was a Conversation scenario and, as shown, produced the largest speed reductions in response to the tunnel. The right panel of the figure compares the Expert and Casual participants' speed reductions at the tunnel (northbound in this case). Participants in the Casual group showed little or no speed decrease (mean = 1.30 km/h) as compared to the Experts' speed reductions during Session 15 (mean = 6.05 km/h decrease). A one-way analysis of variance indicated that the difference between the two groups was statistically reliable [$F_{(1, 18)}$ = 9.826, p < 0.01, $\acute{\eta}_p^2$ = .353]. Previous researchers have noted that drivers entering a tunnel often decrease their vehicle speeds, presumably due to an increase in optical flow rate and inflation of perceived speed (Denton, 1980; Manser and Hancock, 2007; Törnros, 1998). In the context of the current experiment, it is tempting to infer that one effect of extended practice and proceduralization was to reduce the

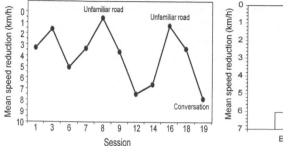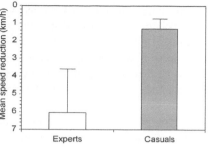

Figure 21.5   **The left panel shows the mean speed reductions in the tunnel across sessions for the Expert group (no tunnel was present during the unfamiliar road scenarios and speed changes shown are for the same section of simulated road). The panel on the right shows the mean speed reductions in the tunnel for the Expert and Casual groups compared for equivalent scenarios (session 15 for the Expert group and session 1 for the Casual group). Error bars show 95 per cent confidence intervals**

participants' explicit attention to the driving task and increase their reliance on implicit perceptual speed cues. Thus, one effect of extended practice was that the participants were more susceptible to the effects of increased optic flow in the tunnel and unconsciously (automatically) reduced their speed. Looking at the conversation scenario (the second scenario of Session 19) the average speed reduction was even greater; perhaps as a result of the participants' attention being drawn away from the driving task for longer periods.

A Tandem Model of Driving

The goal of the pilot experiment described above was to examine whether proceduralized (automatic) driving could be produced with extended practice in a driving simulator. Taken as a whole, the results suggest that it is indeed possible to detect some important changes in driver behaviour with extended practice. Some of the principal changes noted were reductions in speed variability and lane position variability. The participants also displayed increased stereotypy of performance on the visual detection tasks. With practice, targets (Volkswagens) were detected further away, even generalizing to situations outside the experimental sessions. Finally, participants displayed perceptual speed regulation at a tunnel, suggesting that extended practice may reduce drivers' explicit attention and increase their use of implicit perceptual speed cues.

We contend that the driving behaviour described above provides some unique insights into the 'top-down' and 'bottom-up' modes proposed by previous authors (Ranney, 1994; Summala and Räsänen, 2000; Trick et al., 2004). How many, and which of these changes should be considered characteristics of bottom-up or automatic driving is open to debate (which we will address shortly), but it is also important to consider carefully the meaning of the terms 'top down' and 'bottom-up'. Martens and Fox (2007) and others (Borowsky et al., 2008) have referred to stereotyped scanning patterns resulting from practice as being a top-down influence, whereas we would regard the same proceduralized patterns as bottom-up and reserve the term top-down for conscious, active information processing of the driving environment. We have previously used the term 'attentional' and 'perceptual' to differentiate these modes (Charlton, 2004) and other authors have used a range of other terms such as 'open-loop' versus 'closed-loop', 'skill-based' versus 'knowledge-based', 'implicit' versus 'explicit' and 'automatic' versus 'controlled'. We believe that none of these terms adequately reflect the phenomenon at hand, in part because driving performance may not be governed by two dichotomous processing states in an 'either – or' fashion.

As an alternative, we propose a 'Tandem model' of driving. In contrast to a more traditional dichotomy between automatic and controlled modes of driving; we believe that an 'operating process' and 'monitoring process' are engaged in tandem. We conceptualize the operating process as a conscious, intentional search of the environment that requires effort and can be undermined by distractions that also require effortful processing. In contrast, the monitoring process is an unconscious error monitoring system that requires relatively little cognitive effort and continues until an error is detected or it is terminated by a conscious choice. The main function of the monitoring process is to review input with regard to a stored template for performance or information, and when a stimulus indicative of potential failure is detected, the object receives additional activation until the operating process can be brought to bear on the situation and select an appropriate action.

In other words, the two processes work in tandem to guide and maintain driver behaviour. The operating process represents an active, intentional level of task engagement that is required when a driver lacks experience or may be required for an unusual or dynamically changing traffic situation. The monitoring process is continuously engaged with the driving task, its function to compare incoming stimuli to stored representations of previous instances of driving, particularly instances indicative of potential errors or hazards. When the incoming stimuli are congruent with stored representations of familiar or benign situations the monitoring process alone is sufficient to maintain most aspects of the driving task without active attention. When elements of the driving situation do not match stored representations, or when stimuli associated with failed control or potential hazard is detected, the activation of attentional pathways is increased, increasing the likelihood that the items will surface in consciousness and the operating process can be applied to the situation. The process of proceduralization represents a broadening and refinement of the templates (schemata) used by the monitoring process to the point where a wide range of familiar situations and circumstances can be handled with little or no activation of the operating process.

Our conceptualization of two processes working in tandem is similar to Wegner's (1994) account of Ironic processes in self-control of mental states. As described by Wegner, an effortful, intentional operating process searches for mental contents consistent with an intended state and an unconscious monitoring process tests whether the operating process is needed by searching for mental contents inconsistent with the intended state. The idea that transient activation of incoming stimuli by an unconscious monitoring process can trigger a shift to sustained, explicit attention is also a feature of current conceptualizations of selective attention and attentional blindness (Most et al., 2005). Similarly, the suggestion that unconscious activation of incoming stimuli previously associated with negative affective states can unconsciously guide performance and produce greater attentional capture for these stimuli is an integral part of the Somatic-Marker Hypothesis of decision-making (Damasio, 1994, 1996). The Tandem model also appears to be congruent with the concept of a Behavioural Inhibition System (BIS) developed to account for the neuropsychological basis of anxiety. This system is believed to be constantly acting in a monitoring capacity, influencing behaviour when a conflict or threat is detected (Gray, 1976). Animal research has since confirmed the key role of the septo-hippocampal pathway and the hippocampus in detecting mismatches between ongoing behaviour and environmental threats or novelty (for example, Gray and McNaughton, 1983, 2003) and given the hippocampus' role in processing spatial information would appear ideally suited for hazard detection and redirection of behaviour and attention while driving.

According to the Tandem model, the early sessions of the present experiment required a considerable degree of explicit engagement of the participants' operating process, there being relatively little stored information available to enable the monitoring process to guide performance. With experience, enough information was accumulated in the templates to allow the participants to rely increasingly on the unconscious monitoring process to guide moment-to-moment control. We contend that this development was reflected in the participants' progressively lower variability of speed and lane position. Another indication of this change was the participants' speed reductions in the tunnel; instead of consciously monitoring their speeds with the speedometer, drivers in the later sessions unconsciously reduced their speed as a consequence of the monitoring process erroneously 'matching' the increased rate of optic flow produced by the tunnel to cues associated with a higher velocity than was actually the case.

The increasingly stereotyped visual detection patterns noted in the present experiment as well as by Martens and Fox (2007) could be interpreted as an indication that drivers' visual scanning has been left largely to the automatic monitoring process; a circumstance which can lead to attentional blindness for changes to some types of environmental stimuli (Galpin et al., 2009). The progressive improvement in target (Volkswagen) detection, as well as its generalization to situations outside the laboratory, could also be interpreted as evidence of modifications to the templates used by the monitoring process through the process of proceduralization. Rapid detection of important stimuli as well as automatic execution of stored responses may both be important characteristics of the monitoring process.

The Argument for Automaticity

Finally, we need to return to the consideration of whether the changes in driver behaviour observed in the pilot experiment can be deemed to represent automaticity in driving. Two sorts of questions have been raised with regard to previous suggestions that various aspects of driving can be automatic or proceduralized with practice: how can accounts based on habits and automatic behaviours describe error avoidance in a dynamic and open-ended task such as driving; and whether any aspect of driving performance can emulate the rapid and invariant automatic responses demonstrated in other research paradigms (Fuller, 2005; Groeger and Clegg, 1997). As regards the first question, it has been suggested that accounts based on automatic control of driving are 'vulnerable to the implausible requirement to recognize, and learn how to respond safely to, what is a virtually infinite number of roads and traffic scenarios' (Fuller, 2005, p. 463). In response, we would suggest that drivers rely on the automatic monitoring process only in situations that are reasonably familiar (in other words, where incoming stimuli match previously stored representations) and that drivers' schemata or templates may indeed be incomplete with regard to all possible hazards (meaning that hazard detection is a skill acquired gradually through experience). Further, some degree of generalization from stored representations to unfamiliar environments undoubtedly occurs, obviating the presumed requirement that automatic driving processes contain representations of all possible roads and traffic scenarios.

As regards the second issue, the conceptual requirement that any automatic or proceduralized responses show little or no variability in topography or amplitude of execution, it may be the case that such a requirement is not applicable to complex chains of behaviour such as driving. In the case of driving, it may be more appropriate to treat responses of different magnitude or even of somewhat different topography as essentially the same response if they produce functionally equivalent outcomes. This approach, called a macromolar response definition (Logan and Ferraro, 1978), may be a more appropriate way of treating driving and other chained behaviours where quantitatively and topographically different response elements may be required from moment to moment to maintain overall performance on the task. For example, maintaining ones' speed in an automobile may require a range of adjustments in foot pressure from moment to moment, depending on the road geometry and other factors. Should these be considered different responses and rule out any consideration of control by unconscious, automatic habit? When driving is viewed as a chain of ongoing response elements, it is perhaps easier to allow the possibility that the functional outcome (for example, maintaining speed or lane position) is what is being governed automatically, and that some variability in executing individual response elements may be required to achieve that outcome. The increasing stereotypy of speed and lane position in the present experiment is thus consistent with an automaticity account, regardless of the moment to moment levels of accelerator pressure or steering wheel angle.

Conclusion

We believe the Tandem Model described in this chapter complements and reconciles many of the existing models of driver behaviour by suggesting that the monitoring process is responsible for unconsciously detecting potentially important stimuli, and engaging the operating process (after some threshold has been reached), experienced consciously as feelings of driving difficulty (Fuller, 2005), risk (Wilde, 1982), discomfort (Lewis Evans and Rothengatter, 2009), or even a time gap (Galpin et al., 2009). The main point of departure from existing models of driver behaviour is that the monitoring process is viewed as a continuously functioning background process which is used to guide and control driver behaviour in most situations rather than a highly stereotyped motor programme that is triggered in exceptional circumstances or a degraded level of performance that remains when a driver's conscious attention is occupied elsewhere. Clearly much additional work remains on how these processes interact, how frequently and broadly the monitoring process samples different driving situations, and what effect contextual cues or motivational conditions may have on thresholds for attentional activation during performance.

References

Borowsky, A., Shinar, D., and Parmet, Y. (2008). The relation between driving experience and recognition of road signs relative to their locations. *Human Factors, 50*, 173–82.

Brown, I.D. (1962). Measurement of the 'spare mental capacity' of car drivers by a subsidiary auditory task. *Ergonomics, 5*, 247–50.

Charlton, S.G. (2004). Perceptual and attentional effects on drivers' speed selection at curves. *Accident Analysis and Prevention, 36*, 877–84.

Charlton, S.G. (2006). Conspicuity, memorability, comprehension, and priming in road hazard warning signs. *Accident Analysis and Prevention, 38*, 496–506.

Charlton, S.G. (2009). Driving while conversing: Cell phones that distract and passengers who react. *Accident Analysis and Prevention, 41*, 160–73.

Damasio, A.R. (1994). *Descartes Error: Emotion, Reason and the Human Brain.* New York: Putnam.

Damasio, A.R. (1996). The somatic marker hypothesis and the possible functions of the prefrontal cortex. *Philosophical Transactions of the Royal Society of London, Series B, 351*, 1413–20.

Denton, G.G. (1980). The influence of visual pattern on perceived speed. *Perception, 9*, 393–402.

Dingus, T.A., Klauer, S.G., Neale, V.L., Petersen, A., Lee, S.E., Sudweeks, J., Perez, M.A., Hankey, J., Ramsey, D., Gupta, S., Bucher, C., Doerzaph, Z.R., Jermeland, J., and Knipling, R.R. (2006). *The 100-Car Naturalistic Driving Study, Phase II - Results of the 100-Car Field Experiment.* (DOT HS 810 593.) Washington, DC: National Highway Traffic Safety Administration.

Duncan, J., Williams, P., and Brown, I. (1991). Components of driving skill: Experience does not mean expertise. *Ergonomics, 34*, 919–37.

Fitts, P.M., and Posner, M.I. (1967). *Human Performance.* Belmont, CA: Brooks/ Cole.

Fuller, R. (2005). Towards a general theory of driver behaviour. *Accident Analysis and Prevention, 37*, 461–72.

Galpin, A., Underwood, G., and Crundall, D. (2009). Change blindness in driving scenes. *Transportation Research Part F, 12*, 179–85.

Gibson, J.J, and Crooks, L.E. (1938). A theoretical field-analysis of automobile-driving. *The American Journal of Psychology, 11*, 453–71.

Gray, J.A. (1976). The behavioural inhibition system: A possible substrate for anxiety. In M.P. Feldman and A.M. Broadhurst (Eds.), *Theoretical and Experimental Bases of Behaviour Modification*, (pp. 3–41), London: Wiley.

Gray, J.A., and McNaughton, N. (1983). Comparison between the behavioural effect of septal and hippocampal lesions: A review. *Neuroscience and Biobehavioural Reviews, 7*, 119–88.

Gray, J.A., and McNaughton, N. (2003). *The Neuropsychology of Anxiety: An Enquiry into the Functions of the Septo-Hippocampal System* (2nd Ed.). Oxford: Oxford University Press.

Groeger, J.A. (2002). Trafficking in cognition: Applying cognitive psychology to driving. *Transportation Research Part F, 5*, 235–48.

Groeger, J.A., and Clegg, B.A. (1997). Automaticity and driving: Time to change gear conceptually. In T. Rothengatter and E. Carbonell Vaya (Eds.) *Traffic and Transportation Psychology: Theory and Application* (pp. 137–46). Oxford: Pergamon.

Hancock, P.A., Wulf, G., Thom, D., and Fassnacht, P. (1990). Driver workload during differing driving maneuvers. *Accident Analysis and Prevention, 22*, 281–90.

James, W. (1890). *Principles of Psychology.* New York: Holt.

Lee, Y., Lee, J.D., and Boyle, L.N. (2007). Visual attention in driving: The effects of cognitive load and visual disruption. *Human Factors, 49*, 721–33.

Lewis Evans, B., and Rothengatter, T. (2009). Task difficulty, risk, effort and comfort in a simulated driving task: Implications for risk allostasis theory. *Accident Analysis and Prevention, 41*, 1053–63.

Logan, F.A., and Ferraro, D.P. (1978). *Systematic Analyses of Learning and Motivation*. New York: Wiley.

Logan, G.D. (1988). Automaticity, resources, and memory: Theoretical controversies and practical implications. *Human Factors, 30*, 583–98.

Logan, G.D. (1990). Repetition priming and automaticity: Common underlying mechanisms? *Cognitive Psychology, 22*, 1–35.

MacKay, D.G. (1982). The problems of flexibility, fluency, and speed-accuracy trade-off in skilled behaviour. *Psychological Review, 89*, 483–506.

Manser, M.P., and Hancock, P.A. (2007). The influence of perceptual speed regulation on speed perception, choice, and control: Tunnel wall characteristics and influences. *Accident Analysis and Prevention, 39*, 69–78.

Martens, M.H., and Fox, M.R.J. (2007). Do familiarity and expectations change perception? Drivers glances and response to changes. *Transportation Research Part F, 10*, 476–92.

Matthews, R., Legg, S., and Charlton, S. (2003). The effect of cell phone type on drivers' subjective workload during concurrent driving and conversing. *Accident Analysis and Prevention, 35*(4), 451–7.

McRuer, D.T., and Weir, D.H. (1969). Theory of manual vehicular control. *Ergonomics, 12*, 599–633.

Michon, J.A. (1989). Explanatory pitfalls and rule-based driver behaviour. *Accident Analysis and Prevention, 21*, 341–53

Moors, A., and De Houwer, J. (2006). Automaticity: A theoretical and conceptual analysis. *Psychological Bulletin, 132*, 297–326

Morris, G.H. (1981). *George H. Morris Teaches Beginners to Ride: A Clinic for Instructors, Parents and Students*. London: Robert Hale.

Most, S.B., Scholl, B.J., Clifford, E.R., and Simons, D.J. (2005). What you see is what you set: Sustained inattentional blindness and the capture of awareness. *Psychological Review, 112*, 217–42.

Näätänen, R., and Summala, H. (1974). A model for the role of motivational factors in drivers' decision-making. *Accident Analysis and Prevention, 6*, 243–61.

Naparstek, S., and Henik, A. (2010). Count me in! On the automaticity of numerosity processing. *Journal of Experimental Psychology: Learning, Memory, and Cognition, 36*, 1053–59.

Norman, D.A. (1981). Categorization of action slips. *Psychological Review, 88*, 1–15.

Patten, C.J.D., Kircher, A., Östlund, J., and Nilsson L. (2004). Using mobile telephones: Cognitive workload and attention resource allocation. *Accident Analysis and Prevention, 36*, 341–50.

Ranney, T.A. (1994). Models of driving behaviour: A review of their evolution. *Accident Analysis and Prevention, 26*, 733–50

Rawson, K.A., and Middleton, E.L. (2009). Memory-based processing as a mechanism of automaticity in text comprehension. *Journal of Experimental Psychology: Learning, Memory, and Cognition, 35*, 353–70.

Recarte, M.A., and Nunes, L. (2003). Mental workload while driving: Effects on visual search, discrimination, and decision making. *Journal of Experimental Psychology: Applied, 9*, 119–37.

Reid, L.D. (1983). Survey of recent driving steering behavior models suited to accident investigations. *Accident Analysis and Prevention, 15*, 23–40.

Sheridan, T.B. (2004). Driver distraction from a control theory perspective. *Human Factors, 46*, 587–99.

Shiffrin, R.M., and Schneider, W. (1977). Controlled and automatic human information processing. II. Perceptual learning, automatic attending and a general theory. *Psychological Review, 84*, 127–90.

Shinar, D., Meir, M., and Ben-Shoham, I. (1998). How automatic is manual gear shifting? *Human Factors, 40*, 647–54.

Stutts, J., Feaganes, J., Reinfurt, D., Rodgman, E., Hamlett, C., Gish, K., and Staplin, L. (2005). Drivers exposure to distractions in their natural driving environment. *Accident Analysis and Prevention, 37*, 1093–1101.

Summala, H. (1988). Risk control is not risk adjustment: The zero-risk theory of driver behaviour and its implications. *Ergonomics, 31*, 491–506.

Summala, H., and Räsänen, M. (2000). Top-down and bottom-up processes in driver behavior at roundabouts and crossroads. *Transportation Human Factors, 2*, 29–37.

Taylor, D.H. (1964). Drivers' galvanic skin response and the risk of accident. *Ergonomics, 7*, 439–51.

Theeuwes, J. (2000). Commentary on Räsänen and Summala, 'Car drivers' adjustments to cyclists at roundabouts.' *Transportation Human Factors, 2*, 19–22.

Theeuwes, J. (2010). Top-down and bottom-up control of visual selection. *Acta Psychologica, 135*, 77–99.

Törnros, J. (1998). Driving behaviour in a real and a simulated road tunnel: A validation study. *Accident Analysis and Prevention 30*, 497–503.

Törnros, J., and Bolling A. (2006). Mobile phone use: Effects of conversation on mental workload and driving speed in rural and urban environments. *Transportation Research Part F, 9*, 298–306.

Trick, L.M., Enns, J.T., Mills, J., and Vavrik, J. (2004). Paying attention behind the wheel: A framework for studying the role of selective attention in driving. *Theoretical Issues in Ergonomic Science, 5*, 385–424.

Verbruggen, F., and Logan, G.D. (2009). Automaticity of cognitive control: Goal priming in response-inhibition paradigms. *Journal of Experimental Psychology: Learning, Memory, and Cognition, 35*, 1381–88.

Wegner, D.M. (1994). Ironic processes of mental control. *Psychological Review, 101*, 34–52.

Wilde, G.J.S. (1982). The theory of risk homeostasis: Implications for safety and health. *Risk Analysis, 2*, 209–26.

Wilde, G.J.S. (2002). Does risk homeostasis theory have implications for road safety? *British Medical Journal, 324*, 1149–52.

Chapter 22

Road-Rail Level Crossings: Expectations and Behaviour

Jessica Edquist, Christina M. Rudin-Brown and Michael Lenné
Monash University Accident Research Centre

Introduction

Road-rail level crossings are dangerous points in the rail network, as the relatively controlled environment of the railway is opened to the unpredictable behaviour of road users. One potentially dangerous interaction is a mismatch between the warning signals provided, and road users' expectations of warnings. Little is known about how road users expect crossing signals to behave, particularly when the crossing controls do not work as anticipated.

Currently in Australia (as in the USA and many other places), most crossings in built-up areas are equipped with flashing signals that are activated when a train approaches, or when the signal controller cannot determine whether or not a train is approaching (in other words, if there is some problem with the train detection system). When there is no train approaching the crossing, there is no light signal. Some crossings are also equipped with boom barriers, known as gates in the USA, which descend to block the vehicle path after the activation of the flashing lights (to allow large vehicles to complete crossing) but before train arrival. Most rural crossings do not have either lights or booms; instead they are 'controlled' by the placement of signs instructing drivers of the presence of a railway crossing, and that they should stop and/or give way (yield) to trains.

The existing research suggests that users may not understand warning signs at level crossings in the way that designers intend (Mitsopoulos et al., 2002; Richards and Heathington, 1988). This may lead to potentially unsafe behaviour, in terms of both intended and unintended failures to comply with road rules (violations and errors). For example, when warning times (the time between signal activation and train arrival) are greater than drivers expect, the number of crossing violations is higher (Richards and Heathington, 1990). Crossing violations also increase with the frequency of active warning signals in the absence of a train (Wilde et al., 1987; cited in Yeh and Multer, 2007).

A system safety approach requires that such failures are anticipated in the design of the road-rail crossing, rather than placing all the responsibility for safety on the road users (Larsson et al., 2010). For this reason, it is important to investigate road users' understanding of, and behaviour in response to, warning signals.

The present study aimed to investigate differences in behaviour at level crossings with different types of controls (stop signs, flashing lights, traffic lights and boom barriers), using the Advanced Driving Simulator at the Monash University Accident Research Centre. The railway crossings and controls simulated were based on the Australian Standards and characteristics of 55 actual railway crossings. Physically, therefore, the simulation was fairly representative of real-world driving, and relative responses to the different control types should be predictive of real-world behaviour.

Unfortunately there are some parts of the driving environment that it is not possible to replicate. The participants in any driving simulator study are aware that their behaviour is being observed and recorded, and there may be effects of this social context; for example, participants may stop in dilemma-zone situations where in real driving they would continue. There is also the fact that in a driving simulator there is never any danger of physical injury from a crash, which may influence participants' willingness to take risks, perhaps by driving faster than normal and/or continuing through crossings where they would normally have stopped.

It is difficult to determine ahead of time what the combined effects of these social and risk factors will have. For this reason, behavioural and eye-movement data from the simulator were supplemented with a post-drive structured interview. The aim of this part of the study was to elicit participants' knowledge of road rules and expectations of signals, and determine whether any violations of road rules were intentional violations, or errors due to misunderstanding. The results reported here concentrate on these subjective results, and how they match with behaviour in the simulator. Further results from the simulation component of the study can be found in Lenné et al. (2011) and Rudin-Brown et al. (forthcoming).

Method

The behaviour of 52 road users (36 male, mean age 33 years) was examined in a driving simulator experiment. All participants held current drivers' licences, with the average licence tenure being 13 years (range 2–33 years). Participants performed several drives, encountering four railway crossings with various types of signals that required them to stop (flashing lights only, traffic lights, boom barrier with flashing lights and stop sign). At the final crossing, the crossing warning signals indicated that a train was approaching, but no train appeared; for half of the participants this was a traffic light, while for the other half it was flashing lights. After the drives, participants were interviewed about their behaviour in the experiment, and when driving in the real world.

Results

Understanding of Warning Signals at Level Crossings

In general, drivers showed a good understanding of what they were supposed to do at rail crossings. All participants answered that the correct response to a flashing signal at a rail crossing without boom barriers was to stop. Likewise, all participants reported that they would stop for a lowered boom barrier. However, although a descending boom with lights activated also requires a stop, three participants answered that they would continue driving through the crossing if they did not feel they had time to stop before the crossing.

Interestingly, a different three participants reported that they would stop at a signalised railway crossing when the lights were not activated (which indicates that no train is approaching). This may indicate a distrust of railway crossing signals; an assumption that the lack of any light may indicate signal failure; or simply a conservative approach in the face of the extremely costly, although unlikely, possibility of a train approaching without the signals being active.

Traffic Lights at Railway Crossings

A particular focus of the present study was to examine what might happen if traffic lights were introduced at railway crossings. All drivers reported that they would stop for a red traffic light at a railway crossing. In contrast to inactive flashing signals, no drivers reported that they would stop at a railway crossing with a traffic light displaying green. Twenty-nine participants said they would 'proceed' without slowing or checking for trains, compared to six at a normal railway crossing with inactive flashing lights. The majority of drivers would be less conservative about traffic lights at railway crossings than standard signals: in the presence of an amber traffic light, three would 'proceed', 28 would only stop 'if they had time/if it was safe', and 21 would always stop.

Expected Warning Times

The Australian Standard calls for a minimum of 20 seconds warning between the activation of warning signals and the arrival of the train, to allow slow-moving vehicles to clear the tracks. In Victoria, standard practice is to set the signals to activate when the train is 24 seconds from the crossing (the vertical line in Figure 22.1). Figure 22.1 shows that the majority of participants have warning time expectations that are generally in accord with standard practice warning times. However, 25 per cent of participants expected trains to arrive very soon after the warning signals activated.

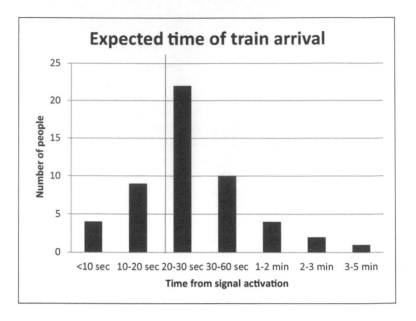

Figure 22.1   Frequency distribution of expected warning times (vertical line indicates standard warning time)

Malfunctioning Warning Signals

Figure 22.2 shows the frequency distribution of the amount of time participants said they would wait at a crossing (or road intersection) without a train (or traffic on the crossroad) in sight before acting.

Forty-five participants reported that after waiting their stated maximum time at a railway crossing, they would cross (after checking for trains); two reported they would find another route, two would call emergency services/the rail authority and ask if there was a signal fault, and three said it would depend on the situation (whether it was possible to see far enough up the line/whether there was an alternative crossing available). All participants reported that they would take the same action at a railway crossing equipped with traffic lights as they would at those equipped with flashing lights. Most participants would also take the same action at a malfunctioning signalised road intersection (although one participant, who reported they would proceed with caution at a malfunctioning rail crossing, reported they would seek another route rather than crossing a road intersection with malfunctioning signals).

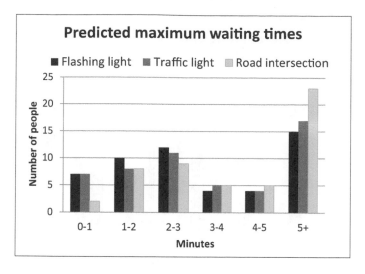

Figure 22.2 **Frequency distribution of predicted maximum waiting times at different types of crossing**

Predicted versus Actual Behaviour

Figure 22.3 shows the frequency distribution of the amount of time participants actually waited when they encountered a crossing with signals flashing but no train arriving in the simulation. The time was recorded from arrival at the crossing, until the participant either spoke to the experimenter, or drove across (in 15 cases). If the participant did not cross or speak within five minutes, the experiment ended. It should be noted that 26 participants saw a malfunctioning flashing light crossing, while the other 26 saw a malfunctioning traffic light crossing, while the frequencies in Figures 22.1 and 22.2 include all 52 participants for each prediction.

For the group that saw malfunctioning flashing lights, the correlation between predicted and actual wait was .33 ($p > .9$). However, for the group that were presented with malfunctioning traffic lights, the correlation was .65 ($p < .001$). This result is unexpected, as both groups predicted they would wait similar amounts of time at flashing lights (2.5 minutes for the malfunctioning flashing light group versus 2.8 minutes for the malfunctioning traffic light group), at traffic lights (both 2.8 minutes), and at road intersections (3.2 versus 3.5 minutes). Interestingly, expected warning time correlated moderately with predicted waits at both flashing lights ($r = .34$) and traffic lights ($r = .31$), but with neither actual wait ($r = .18$ at flashing light, .01 at traffic light).

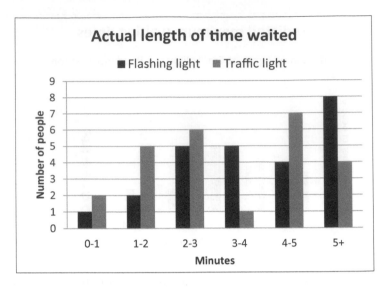

Figure 22.3   Frequency distribution of actual times waited at different crossing types

Crossing Violations

A logistic regression was undertaken to determine whether a combination of demographic variables, expected warning time and predicted wait time could predict those who would violate the final crossing with the malfunctioning signals. No variables increased the predictive success of the equation beyond the initial model (which predicted no violations).

The demographic characteristics of the participants were examined separately in four groups (see Table 22.1): those who never violated a crossing signal, those who violated a crossing signal (boom barrier, flashing light, traffic light or stop sign – mostly stop signs) during the drive, those who violated the crossing with the malfunctioning signal (after coming to a stop and waiting for some time) and those who violated both the malfunctioning crossing and another crossing during the drive.

In general, the characteristics of those who violated crossing signals were fairly similar to those who did not violate crossing signals, although there are a few notable trends (statistical tests have not been performed due to the small size of some groups). The 'never violate' group appeared to be slightly older and more experienced. There were more males in the group who only violated the normal crossing signals, and fewer in the group who never violated crossings. The group who only violated normal signals contained more drivers who were required to wear some form of optical correction (glasses or contact lenses) while driving, which may indicate that these drivers had difficulty perceiving signs/

signals in the simulator. The group who violated both normal and malfunctioning signals appeared to drive somewhat more (when measured in hours per week spent driving – many participants had difficulty estimating how many kilometres per week they drove, so the latter figures may not be as reliable).

Table 22.1   Demographic characteristics of crossing violators

	N	% male	Age	Age at licence	Years licensed	# lenses required	hrs/wk	km/wk	Environment driven most
None	14	57%	35.1	18.3	16.9	1	9.8	253	residential
Malfunction	6	67%	31.3	18.8	12.5	1	9.5	335	metropolitan
Normal	23	78%	32.1	19.2	13.0	8	7.8	192	metropolitan
Both	9	67%	31.8	19.7	12.1	2	10.8	331	residential
Total	*52*	*69%*	*32.8*	*19.0*	*13.8*	*12*	*9.1*	*249*	*residential*

Discussion

In general, drivers showed a good understanding of what they were supposed to do at rail crossings. If traffic lights were to be introduced at crossings, the evidence of the present study indicates that they would also be understood. Participants' responses indicated that they would be more likely to cross without slowing and checking for trains when presented with a green traffic signal at a railway crossing than when presented with a flashing light array with no lights flashing. This may be perceived as dangerous (if the signals are not reliable) or as a benefit (as the increased predictability of traffic movement may prevent vehicle-vehicle crashes and improve traffic flow).

A minority of participants in this study expected trains to arrive no more than 20 seconds after warning signals activate. Road users who expect short warning times may lose confidence in the accuracy of the signals when the train fails to arrive quickly, and may attempt to cross despite the signals, leading to a high-risk incident with a low safety margin between the vehicle clearing the crossing and the train entering the crossing, or potentially a crash.

When participants experienced an activated warning signal that never eventuated in a train, about one-third of the group crossed against the signal, despite the knowledge that they were being observed. The data collected in the present study could not differentiate those participants from others. However there were indications that gender, age/experience and driving exposure may affect these tendencies.

The present study is consistent with previous research showing that people will disregard rules when they believe the violation is unlikely to result in a large risk (for example, Nyssen and Cote, 2010). Participants in our study stressed that they would check thoroughly for trains before violating a crossing, but most would disregard the road rules rather than inconvenience themselves by taking another route.

Self-reported intentions to perform a behaviour have been found to be highly correlated with actual behaviour (Ajzen, 1991), and are often used as a surrogate measure for actual behaviour. In the present study, participants' descriptions of how they would behave (in other words, their intentions to perform a behaviour) at a crossing with malfunctioning signals did not always match their behaviour when they encountered this situation in the simulation. This has implications for the theory of planned behaviour and similar theories that attempt to predict behaviour from intentions or motivations. The present study is of course limited by the fact that behaviours were collected in a simulated road environment, rather than a real road environment. However, the results suggest that caution should be used when interpreting studies based solely on self-reported behaviour or behavioural intentions without any comparative data.

References

Ajzen, I. (1991). The theory of planned behavior. *Organizational Behavior and Human Decision Processes, 50*, 179–211.

Larsson, P., Dekker, S.W.A., and Tingvall, C. (2010). The need for a systems theory approach to road safety. *Safety Science, 48*, 1167–74.

Lenné, M., Rudin-Brown, C.M., Navarro, J., Edquist, J., and Trotter, M. (2011). Driver behaviour at rail level crossings: Responses to flashing lights, traffic signals and stop signs in simulated rural driving. *Applied Ergonomics, 42*, 548–54.

Mitsopoulos, E., Regan, M.A., Triggs, T.J., Wigglesworth, E., and Tomasevic, N. (2002). Do pictogram signs increase safety at passive crossings? A simulator experiment. Paper presented at the *7th International Symposium on Railroad-Highway Grade Crossing Research and Safety*, 2002, Melbourne, Victoria, Australia.

Nyssen, A-S., and Cote, V. (2010). Motivational mechanisms at the origin of control task violations: An analytical case study in the pharmaceutical industry. *Ergonomics, 53*, 1076–1084.

Richards, S.H., and Heathington, K.W. (1988). Motorist understanding of railroad-highway grade crossing traffic control devices and associated traffic laws. *Transportation Research Record, 1160*, 52–9.

Richards, S.H., and Heathington, K.W. (1990). Assessment of warning time needs at railroad-highway grade crossings with active traffic control. *Transportation Research Record, 1254*, 72–84.

Rudin-Brown, C.M., Lenné, M.G., Edquist, J., Trotter, M., Navarro, J., and Tomasevic, N. (forthcoming). Traffic light versus boom barrier controls at road-rail level crossings: A simulator study. *Accident Analysis and Prevention.*
Yeh, M., and Multer, J. (2007). Traffic control devices and barrier systems at grade crossings: Literature review. *Transportation Research Record, 2030,* 69–75.

Chapter 23

Stochastic Changes in Driver Reaction Time with Arousal State

Takahiro Yoshioka
Graduate School of Information Science and Electric Engineering, Kyushu University

Shuji Mori
Faculty of Information Science and Electric Engineering, Kyushu University

Yuji Matsuki
Faculty of Engineering, Fukuoka Institute of Technology

Osamu Uekusa
Domestic Sales Strategy Promotion Department, UD Trucks Corp.

Introduction

The purpose of this study is to examine how reaction time (RT) distributions of an automobile driver change, in a stochastic manner, with the driver's arousal state. Most of the assessment indices proposed thus far (for example, headway time and time-to-collision) are computed solely from physical parameters, such as driving velocity and distance from a lead car (Evans, 1991; Hayward, 1972; Vogel, 2003). Although it is well known that human factors like arousal state are strongly related to automobile accidents (Dinges, 1995), human factors were not included in these indices. It is also well known that RTs during driving are strongly dependent on the driver's arousal state (Wlodarczyk et al., 2005), but how the RT distribution changes with the arousal level has yet to be determined. Such stochastic properties of RTs are crucial for computing the probability of collision (POC), an index of the likelihood of an accident at a given moment (Matsuki et al., 2004).

In our previous investigation (Yoshioka et al., 2009), we measured RTs in simplistic driving conditions. To ascertain the stochastic properties of RT data, we modelled RT distributions using an ex-Gaussian function, which has been found to fit RT data well (Luce, 1986). As an index of the participant's arousal state during the experiment, we used eye-opening ratio (EOR), which is the distance between the upper and the lower eyelids at a given moment relative to the distance exhibited under a high arousal state (Mineyama et al., 2006; Yoshioka et al., 2009). The results showed that the RT distributions changed in shape, with longer means and larger variances, as EOR decreased. Probability of collision, computed from the distributions, became higher (meaning that the likelihood of an accident increased) as EOR decreased.

In the present study, we measured RTs in a more realistic setting, where the participants drove a simulator and hit the brake pedal in response to the sudden stopping of a lead car. Then we computed the POC at low and high EOR values from the data, for each participant.

Experiment

Participants

Five females and five males, who ranged in age from 21 to 25 years old, participated in the study. One of them was the first author. All of the participants had a Japanese driver's licence at the time of participation and none of the participants had difficulty in seeing the stimuli used in this experiment.

Apparatus

A custom-made driving simulator was used. The simulator consisted of a driver's seat with a handle (not used here), brake and accelerator pedals, and a 42-inch full-colour display (Toshiba, Regza 42C3500) that presented computer graphics simulating the front view of a car. The graphics were created using a personal computer (Dell, Precision T7400), an application programming interface for multimedia processing (Microsoft Direct X 9.0), and image editing software (e-frontier Shade 10 and Adobe Photoshop CS3). This computer also controlled the experimental timing and RT measurement with an optical time counter (Interface, LPC-630101). The velocity of the car in the graphics was controlled using the brake and accelerator pedals. A second computer (custom-made) received data from the pedals and transmitted them via a gigabit Ethernet LAN to the above-mentioned simulator-control computer. Using this data, the simulator-control computer calculated the positions of all materials in the graphics for the next frame. For image processing the driver's eye, which was used for measuring EOR, we used a video camera (Sony, HDR-SR1) with 56 infrared LEDs, which enabled images to be obtained in the dark. The images were captured at 30 Hz with a frame grabber (Cyber Optics, PXC200) on a computer (DELL, Precision 690). The spatial resolution of the images was 1024×576 pixels.

Stimuli

The front view of the simulator showed two straight lanes, where the driver's car moved in the left lane, with a black car driving ahead in the same lane. The participant controlled their car's velocity using the accelerator and brake pedals, but they were not allowed to change lane or direction. The lead car moved forward at 60 km/h, except for when it slowed down suddenly. The stimulus for the participant to respond to was the brake lights on the rear of the lead car. When the brake lights were activated and the lead car slowed down suddenly, the participant was required to stop their car by pressing down on the brake pedal to avoid collision. The timing of the brake lights being turned on followed an exponential distribution of time from the previous turning on of the brake lights, with a mean of 30 seconds. Exponential distributions have previously been used to set foreperiods for stimulus presentation in simple RT experiments (Luce, 1986). In this experiment, there were additional constraints on the stimulus presentation in that the brake lights were only turned on when the distance between the two cars was between five and 30 metres and the driver was pressing on the accelerator pedal. The luminance of the brake lights when they were on and off was 78.8 and 8.62 cd/m², respectively. These values adhere to the motor vehicle safety standard in Japan.

Procedure

Participants were seated in the driving simulator. Before the experiment, they were instructed to keep looking at the rear of the lead car shown in the front view of the simulator display while controlling the following car's velocity (using the accelerator) and to press the brake pedal as soon as the brake lights of the lead car were activated. The participants were given a few practice trials before proceeding to the experimental session. In each session, the lead car started to move and accelerated to reach a constant speed of 60 km/h and the participants controlled the following car to keep a stable distance with the lead car until the brake lights were turned on. The time between the brake lights being turned on and when a participant pressed the brake pedal was measured as the RT with millisecond precision. The lead car started to move again after the participant's car had completely stopped. Each session consisted of 20 stimulus presentations and thus 20 RT measurements. Each session lasted 20 to 40 minutes and each participant completed 15 sessions over one to three weeks, depending on their schedule.

Data Analysis

To measure eyelid positions accurately in real time, a small reflection sheet (a circle of 8 mm in diameter) was attached to the participant slightly below the lower eyelid of the left eye, so that the lower eyelid position in the captured image was defined relative to the centre of the attached sheet. The edge line of the upper eyelid was detected from the luminance gradient of the upper eyelid. The upper eyelid position was defined as the intersection of a vertical line drawn from the lower eyelid position with the edge line of the upper eyelid. The distance between lower and upper eyelid positions was computed and taken as the vertical width of eye opening.

In the present study, momentary eye-opening width was the average value of eye opening over a one-second time window. The eye opening at a high arousal level was the average value of eye opening from 10 to 20 seconds after the beginning of each experimental session. We classified EOR values into two categories, low (lower than 85 per cent) and high (higher than 85 per cent), and computed POC from RT data separately for these categories. A previous study has shown that this classification yields sufficient numbers of RT data for each EOR category (Mineyama et al., 2006). Our preliminary investigation also showed that these two categories of EOR values are correlated with verbal reports of drowsiness during a RT experiment (Yoshioka et al., 2009).

An ex-Gaussian distribution (Luce, 1976) is expressed by the following equation:

$$f(x) = \lambda \exp(-\lambda t) \exp\left[\lambda\mu + \frac{1}{2}\lambda^2\sigma^2\right] \Phi\left(\frac{t - \mu - \lambda\sigma^2}{\sigma}\right)$$

where $\Phi(x)$ is a standard normal cumulative distribution function, μ and σ are the mean and the variance of the Gaussian distribution, respectively, and λ is the inverse of the mean of the exponential distribution. Ex-Gaussian is the convolution of exponential and Gaussian density functions, and it has been found to be effective in approximating simple RT data (Luce, 1986).

In the present study, fitting to the ex-Gaussian distribution was performed separately for both high and low EOR values for each participant. Firstly, the probability density was obtained from each set of RT data by Gaussian kernel estimation (Silverman, 1986; Yoshioka et al., 2010). Next, the ex-Gaussian distribution was fitted to the density. Best-fitting parameters for ex-Gaussian were chosen by the function called 'NonlinearFit', which appears in a formula manipulation application (Wolfram Mathematica 5.2).

Results

Figure 23.1 shows estimated probability density and fitted ex-Gaussian distributions for one participant. The circles show the probability density obtained by Gaussian kernel estimation. The lines show the best-fit ex-Gaussian functions fitted to the probability density. For either EOR category, the ex-Gaussian function captured the main features of the shape of the probability density. Other participants showed similar results.

Figure 23.2 shows the averages of the means of the row RT data and of the means of the fitted ex-Gaussian for all but one participant (S10), who gave much longer RTs for low EOR (see also Figure 23.3). The bars show standard errors of the means. As is clearly seen from the figure, the means of the fitted ex-Gaussian

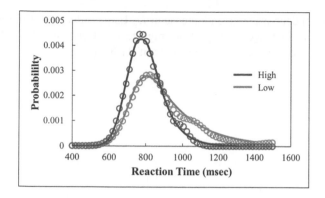

Figure 23.1 Results of one participant (S1)

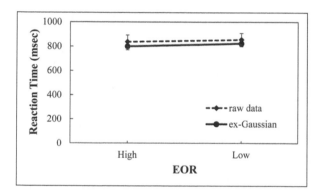

Figure 23.2 Average of the means of the raw RT data and the means of the fitted ex-Gaussian distributions of all participants

distributions are very close to the corresponding means of the RT data. A 2 (Data Type: raw data, fitted ex-Gaussian) × 2 (EOR: high, low) ANOVA showed no significant main effect for Data Type, $F(1, 16) = 0.91$, or EOR, $F(1, 16) = 1.23$, and no significant interaction between the two factors, $F(1,16) = 0.03$. This confirms the high accuracy of the fitting of ex-Gaussian distributions to RT data.

We computed POC separately for high and low EOR for each participant from their best-fit ex-Gaussian functions. For details of the computation of the POC, see Yoshioka et al. (2010). Figure 23.3 shows the results of all 10 participants as a function of the following distance from the lead car, D_f. The velocity of the following car at the moment of the sudden stop of the lead car was set at 60 km/h. With one exception, the individual POC curves share common features: POC is 1.0 when D_f is smaller than 30 m, and it drops sharply to zero as D_f increases from 30 m to more than 40 m. The exception is the results of S10 for low EOR, which show a gradual decrease with an increase in D_f. Large individual differences were noted in terms of change in shape from high to low EOR, and these differences largely reflect the differences in the change of RT histograms.

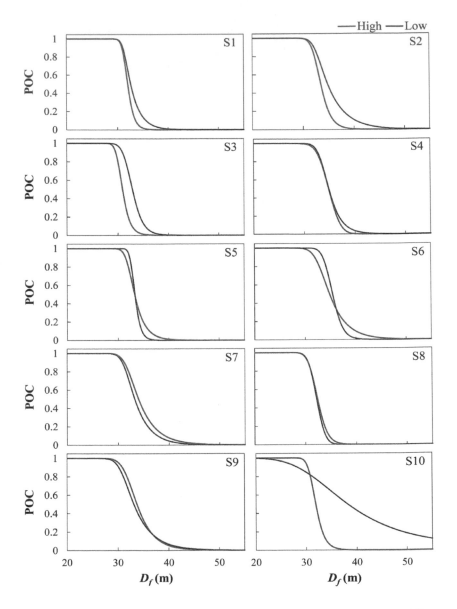

Figure 23.3   Individual representations of POC for high and low EOR, as a function of following distance D_f

Discussion

As shown by the quantitative comparisons of the RT data and their corresponding ex-Gaussian functions (Figure 23.2), the fitting of ex-Gaussian functions to the histograms was quite successful, which suggests that the ex-Gaussian function would be appropriate as a model of RT distribution during automobile driving. However, separate computation of POC for high and low EOR exhibits problems. There are large variations in the shapes of the POC curves among participants. It was anticipated, before the experiment, that POC would be higher for low EOR than for high EOR, that is, the likelihood of accident involvement would be higher at low arousal states than at high arousal states. Although some participants showed such results (S1, S2 and S3), other participants showed indistinguishable changes in POC from high to low EOR (S4 and S8), or even higher POC for high EOR than for low EOR (S7 and S9). Furthermore, the two POC curves of several participants intersected with each other (S5, S6, and S10).

The primary reason for the individual variation and unexpected patterns of POC is that the number of RT data points classified into either EOR category was too small to enable a reliable estimation of RT distribution; in many cases, the numbers for low EOR were smaller than for high EOR. For several participants, there were less than 30 RTs for low EOR (S3, S6, S7 and S10). The small sample size leads to errors in the estimation of the true distribution. As the number of data points for either RT category becomes small, it is more likely that the Gaussian kernel estimates of RT distribution are not accurate, resulting in aberrant patterns of POC.

References

Dinges, D.F. (1995). An overview of sleepiness and accidents. *Journal of Sleep Research 4*, 4–14.

Evans, L. (1991). *Traffic safety and the driver*. New York: Van Norstrand Reinhold.

Hayward, C.J. (1972). Near miss determination through use of a scale of danger. *Highway Research Board*, 24–34.

Luce, D.R. (1986). *Response times: Their role in inferring elementary mental organization*. New York: Oxford University Press.

Matsuki, Y., Matsunaga, K., and Shidoji, K. (2004). On the development of a real-time evaluation system for safe driving. In Proceedings of the *International Conference on Application of Information and Communication Technology in Transport Systems in Developing Countries*, CD-ROM.

Mineyama, T., Matsuki, Y., Shidoji, K., and Uekusa, O. (2006). Estimating reaction time by analysing eyelid image. *IEICE Technical Report*, 106(266), 11–14.

Silverman, W.B. (1986). *Density estimation for statistics and data analysis*. London: Chapman and Hall.

Vogel, K. (2003). A comparison of headway and time to collision as safety indicators. *Accident Analysis and Prevention, 35*, 427–33.

Wlodarczyk, D., Jaskowski, P., and Nowik, A. (2005). Influence of sleep deprivation and auditory intensity on reaction time and response force. *Perceptual and Motor Skills, 101*(3), 949–960.

Yoshioka, T., Mori, S., Matsuki, T., and Uekusa, O. (2009). From RT to POC: Proposal for computation of probability of automobile accidents from empirical reaction time distribution. *Proceedings of the 2nd International Multi-Conference on Engineering and Technological Innovation, 2*, 7–10.

Yoshioka, T., Mori, S., Matsuki, T., and Uekusa, O. (2010). Estimating probability of automobile accident from driver's reaction time under different arousal states. In Proceedings of *17th ITS. World Congress*, CD-ROM.

PART VI
Technology in Vehicles and
User Acceptance

Chapter 24

Using Local Road Features and Participatory Design for Self-Explaining Roads[1]

Samuel G. Charlton

Traffic and Road Safety Research Group, University of Waikato, New Zealand

Introduction

This chapter describes two recent projects undertaken to establish self-explaining road (SER) hierarchies on existing streets in an urban and a suburban area. In order to be cost-effective and be appropriate to local considerations the projects made maximum use of road characteristics already present (in other words, endemic features) and relied heavily on community participation during the design process.

Self-Explaining Roads

In recent years there has been a move to use the visual features of roads to control speeds instead of relying solely on physical barriers or enforcement. When the intended functions of roads are clear to users and consistently reflected in their design they can be said to be self-explaining (Theeuwes, 1998; Theeuwes and Godthelp, 1995; Rothengatter, 1999; Weller et al., 2008). The SER approach has its roots in cognitive psychology and attempts to improve road safety via two complementary avenues. The first is to identify and use road designs that promote desirable driver behaviour. Perceptual properties such as road markings, delineated lane width, and roadside objects can function as affordances that serve as built-in instructions and guide driver behaviour, either implicitly or explicitly (Charlton, 2004, 2007; Elliot et al., 2003; Lewis Evans and Charlton, 2006; Weller et al., 2008). A second aspect of the SER approach is to help establish mental schemata and scripts, memory representations that will allow road users to easily categorize the type of road on which they are travelling and behave accordingly (Theeuwes and Godthelp, 1995). When the visual features of roads are applied consistently

---

1 Paper presented at: International Congress of Applied Psychology, 12 July 2010. Originally titled *Improved Safety and Reduced Speeds Resulting from a Self-explaining Roads Process.*

within a hierarchy of road types, drivers will be more likely to form schemata that automatically evoke the desired expectations and driving behaviours. Thus, the SER approach both advocates the use of road designs that build on existing affordances around road use and employs road designs consistently to assist drivers in forming appropriate schemata for various categories of road (including the desired speed), promoting successful categorization, and as a result, the correct behaviour for that road.

Fundamental to establishing SERs is ensuring that there is a clear hierarchy of roads based on their intended functions and a consistent set of road designs that are appropriate to those functions. At the level of implementation, however, there is the practical question of what road designs should be used to reflect the functional hierarchy. A variety of methods for choosing category-defining road features have been proposed. For example, Kaptein and Claessens (1998) demonstrated the use of a picture-sorting task to select road characteristics that would lead to maximum visual discriminability between different road categories. Wegman et al. (2005) described the use of a formal functional requirements analysis to determine the designs for each road category that would meet the three SER principles of functional use, homogeneous use, and predictable use. Herrstedt (2006) reported the use of a standardized catalogue of treatments compiled from researcher and practitioner advice. Goldenbeld and van Schagen (2007) used a survey technique to determine road characteristics that minimize the difference between drivers' ratings of preferred speed and perceived safe speed and select road features that make posted speeds 'credible'. Aarts and Davidse (2007) used a driving simulator to verify whether the 'essential recognizability characteristics' of different road classes conformed to the expectations of road users.

In many cases, however, the goal is to apply SER design principles to existing roads. When attempting to apply a design hierarchy in established neighbourhoods there is a range of constraints on the applicability of methods one can use. There are residents with a sense of ownership over the roads and the way they look, there are road users with established habits, and constraints posed by the existing infrastructure as to what can be achieved at reasonable cost. To address these constraints (be sensitive to existing aesthetics, build on existing habits, and be cost effective) we employed endemic features (already present) and actively involved local residents in identifying the desired road functions and selecting road designs appropriate and acceptable for each level of the hierarchy (Charlton et al., 2010). This chapter briefly describes two projects employing this approach, one conducted through to the completion of a conceptual design, the other carried through to implementation and monitoring of its effectiveness.

Project 1 – An Urban Neighbourhood

The study area for this project was an established urban neighbourhood that contained a mix of private residences, small shops, schools, and churches, and was selected in consultation with representatives from a number of national and local transport agencies. The study area was divided into two equivalent sections, one to receive SER treatments and the other to serve as a control section. Each of these sections comprised an area of approximately 1.3 km² and contained approximately 14 km of public roads. The number of vehicle movements per day on these roads ranged from an average of 146 vehicles per day to nearly 20,000 vehicles per day.

Identifying a Road Hierarchy

Although the area was an established urban area, there was no clear consensus as to the existing hierarchy of streets in the area. Public maps, local council plans, and road maintenance documents indicated several divergent classifications of the roads in the study area. Some roads designated as collectors in one source were designated as local roads by another and some roads listed as regional arterials in other sources were shown as collector roads in another. In order to resolve these different points of view we conducted two sorts of surveys. The first was a set of speed surveys collected by placing tube counters on the roads in the study area, the result of which were data showing the traffic volumes and distribution of speeds over seven days. The second survey was a road and travel survey completed by 230 local area residents. The survey requested information on routes and modes of travel to different destinations throughout the day and was distributed to residents via area schools, community groups, and at local shops.

Combining the survey responses from the local residents and the traffic counts we were able to identify a three-level road hierarchy for the study area. The traffic volumes in the study area fell into three readily identifiable clusters: regional arterials (that ranged from 13,500 to 19,750 vehicles per day); collector roads (5,000 to 8,000 vehicles per day), and local (access) roads (that ranged from 100 to 2,000 vehicles per day). The resulting three-level road hierarchy also provided a reasonable match to a proposed national road hierarchy independently derived from national traffic volumes and road asset management documents (Macbeth, 2007). Although posted speed limits were the same for all roads in the study area (50 km/h), average speeds, 85th percentile speeds, and crash rates varied widely within and across each level in the hierarchy. In consultation with planning staff from the local council we also identified some 'design-to' or idealized speeds for each level of the new hierarchy (although legal speed limits would remain at 50 km/h). A design speed of 30 km/h was chosen for local roads, 40 km/h was selected for the collector roads, and 50 km/h for the regional arterials.

Analysis of Road Features

At the same time as the traffic volumes and speeds were surveyed, a range of road features were measured and tabulated for each road in the treatment area, including road width, road delineation and signage, sight distances, landscaping, location of parking, footpaths, and road reserve widths. The physical, visual, and functional (speed and traffic volume) characteristics of each road in the study area were then used to identify (1) roads that possessed the speeds and traffic volumes that fit with their level in the new hierarchy (functional exemplars), (2) roads with speed, safety, or volumetric problems (functional outliers), and (3) roads with design properties that differed from their designated function (design outliers). Functional exemplars were identified at each level of the road hierarchy and the road characteristics that differentiated them from the outliers were identified. Analysis of these road characteristics was also aimed at identifying design features that could be used to emphasize visual differences between each level of the three-level road hierarchy. As shown in Figure 24.1, visually there was very little to differentiate roads carrying little or no traffic (local roads) and collector roads carrying high volumes of traffic.

Local road: Average of 1,182 vehicles per day   Collector: Average of 5,843 vehicles per day   Arterial: Average of 19,749 vehicles per day
Average speed 44.7 km/h   Average speed 50.9 km/h   Average speed 54.0 km/h
85th percentile speed 54.4 km/h 85th percentile speed 58.0 km/h 85th percentile speed 61.0 km/h

Figure 24.1   Roads in the study area before treatment

Development of Design Template

The next step in the process was to develop a conceptual design template for the study area that would reinforce the desired functions at each level in the road hierarchy (local, collector, and arterial roads). The visual features of the functional exemplars were used to identify road characteristics that could be applied to the streets identified as problem areas (functional outliers and design outliers), while giving the entire study area a consistent look and feel. One feature present in many of the local (access) roads that were working well was restricted forward visibility resulting from landscaping and road geometry. In contrast, the local roads classified as functional outliers had very good forward visibility, in some cases exceeding 500 m. Inasmuch as long forward visibility was also a characteristic of the collector and regional arterial roads in the area, we decided to use restricted forward visibility as a category-defining feature of the local road category. For local roads where this feature was lacking we proposed to use a combination of trees planted in the centre of the road and landscaped 'community islands' placed at the curb sides to limit forward visibility.

For the collector-road category, we decided to use a high standard of road delineation as a category-defining feature. Centrelines and edgelines added to the collector roads that lacked them and cycle lanes, pedestrian crossings, and low-lying landscaped medians with pedestrian refuges were included in the design template for collector roads. These road delineation features were intended to clarify each road user group's place on the road and make the collector roads clearly distinct from local roads (as shown in Figure 24.2; local roads on the left, collector roads on the right). Conversely, we planned to remove any existing road delineation and other road markings from the roads in the local category. Because of several operational issues associated with the regional arterials, they were excluded from the subsequent design and evaluation stages of the study.

Figure 24.2   SER conceptual designs for the local road category (left) and collector road category (right)

Community Participation in the Designs

Extensive consultation with the community occurred during the design process. Artwork containing prototypes of candidate designs were presented to the area residents at a series of community workshops to get resident feedback and input (See Figure 24.3).

In addition, many small group meetings and individual appointments with community groups were held to ensure a broad representation of resident and road user feedback. Following an iterative process of design-budget-feedback-redesign, the SER template and designs were presented to the community in an open day event organized on-site in the treatment area to show local residents examples of the location and type of treatments that were going to be placed on their roads.

Implementation and Evaluation

In total, approximately 11 km of local and collector roads were modified within the study area over a period of approximately four months. Examples of the completed treatments are shown in Figure 24.4. Three months after completion of all construction, speed and traffic data were collected at eight roads by means of tube counters placed in the same locations as the pre-treatment speed data. Three of the measurement sites were on treated local roads, three were on treated collector roads, and one local road and one collector road were selected from the control area for comparison purposes. The tube counters remained in place for seven days at each location.

Figure 24.3   Public consultation with residents, early in the design process (left) and on location prior to construction (centre and right)

Figure 24.4   Completed SER treatments: local road (top left), collector road (top right), mountable roundabout at a local road intersection (lower left), and a local road gateway (lower right)

Figure 24.5 shows the mean speeds for the four local roads before and after installation of the treatments. As the figure shows, large reductions in speed were observed to occur for the three local roads in the treatment area, but not for the local road in the matched control (no treatment) area. Mean speeds on the treated local roads declined from 44.39 km/h to 29.62 km/h and eighty-fifth percentile speeds decreased from 54.29 km/h to 36.71 km/h. An analysis of variance comparing the before and after speeds across the three treated local road measurement locations indicated a significant reduction in observed speeds [$p < 0.001$, $\acute{\eta}_p^2 = 0.291$]. Small but statistically significant reductions in mean speeds, ranging from 1.50 to 2.57 km/h, were also observed on the collector roads in the treatment area [$p < 0.001$, $\acute{\eta}_p^2 = 0.024$].

Figure 24.6 shows another aspect of the changes in drivers' speeds on local and collector roads before and after installation of the SER treatments. There was considerable overlap between the distributions of drivers' speeds for local and collector roads prior to treatment. Following installation of the treatment not only were the mean speeds on local roads lower than before, the speed variability for all of the streets was lower as well, with the net effect of producing two substantially different speed profiles for the two road categories.

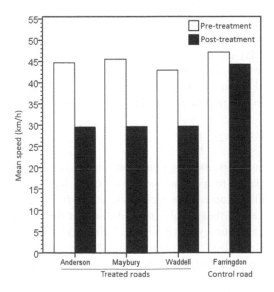

Figure 24.5   Mean speeds on local roads before and after SER treatment. The three roads at left are from the treatment area, and at right one untreated site from the control area

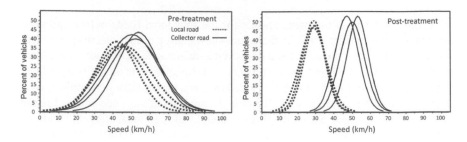

Figure 24.6   Distribution of speeds on local and collector roads before treatment (left) and after SER treatment (right)

Project 2 – A Suburban Community

The study area for the second project was in a predominantly suburban/rural area and was selected in order to apply the SER methods developed in the previous project to an area with different road and traffic characteristics. The area selected for study was a geographically discrete suburban community of approximately 1.6 km² and contained approximately 10 km of public roads. As with the previous project, the study area was an established community with little differentiation in the visual appearance of the streets. Traffic volumes in the area ranged from lows of 21 vehicles per day on some neighbourhood streets to 153 vehicles per day on others with volumes of 1,244 vehicles per day on the arterial road that bisected the community. Posted speeds for all of these roads was 50 km/h but measured mean speeds ranged from 41 km/h to over 65 km/h on sections of the main road.

Public Perceptions Survey

Although there was a history of speed problems on some roads there was also a variety of complaints to local authorities on other issues including pedestrian safety, use of mobility scooters and cycles, and school bus stops. In order to gather more information about these problems and discover the suburban hierarchy of roads, a road and travel survey was mailed to 1,000 local residents. The survey was completed and returned by 168 residents. The responses to the survey indicated a range of specific problems ranging from the need for cycle lanes, safer pedestrian crossings, poor access to shops and schools, uneven footpaths, and speeding cars and trucks. A large number of comments mentioned difficulties posed by the busy arterial road that created barriers to walking and cycling and separated different activities within the community. When asked to rank the transport modes that most needed improvement in the community, the residents ranked cycling as most problematic, followed by walking, mobility scooters, cars, and other (including public transport and skateboards).

The results of the travel survey showed that residents often used circuitous routes through the community to access schools, shops, recreational facilities, and other residences, apparently to avoid or minimize their use of particular streets and intersections. The travel survey also showed that most of the roads were used for multiple functions, in other words there was no clear hierarchy of access, collector, and arterial roads. The survey results informed the focus of the subsequent design activities in the sense that they included not just speed management but also improving walking, cycling, and linkages with public transport. It was hoped that by delineating some preferred routes for specific travel modes and destinations (and providing any infrastructure needed to support those modes) the project would be able to improve connectivity and mobility in the community; increase safety for pedestrians and cyclists; and contribute to a reduction in environmental emissions by reducing the number of short-distance car trips.

Participatory Design Workshops

In order to make maximum use of local expertise in finding design solutions to the problems described in the survey we decided to increase the amount of community involvement through a series of participatory design workshops. Participatory design is a process that involves eliciting the knowledge, ideas, and opinions of users throughout the stages of design (Maarttola and Saariluoma, 2002; Reich et al., 1996; Tang and Waters, 2005). Participatory design techniques have been used in a variety of contexts since the 1970s but perhaps have enjoyed the greatest use in the design of software and human–computer interfaces (Floyd et al., 1989). One of the main goals of participatory design is to elicit the tacit or implicit knowledge of users, knowledge that designers may lack to the extent that they are not users of the system or artefact being designed. Participatory design has enjoyed some success in transport planning and urban design (Hardie, 1988; Maarttola and Saariluoma, 2002; Tang and Waters, 2005), and so for the present project we decided to employ participatory design techniques to use the residents' knowledge about the transport issues in their community.

One approach to eliciting users' knowledge is through the use of exploratory design games (Brandt, 2006). With that in mind, we set about organizing a series of participatory design works that would be enjoyable for the participants and produce a wide variety of information about transportation problems and possible design solutions. Two participatory design workshops were held with participants invited from a list of survey respondents who indicated that they would be willing to be involved in future meetings about transportation issues. At the workshops the participants were given a short presentation on the results of the survey and the travel patterns in the community, the vehicle speeds and traffic counts recorded on roads in the area, and some background on the SER approach. Following the presentation the participants were divided into two groups of approximately 10 to 12 residents each. A traffic engineer from the local council was assigned to each group to listen to the issues raised by the residents and assist them by answering questions. Other residents were free to observe the activities of both groups without taking a direct part in the planning activities.

Each group was provided with enlarged maps of the area, each group's maps focused on a different part of the community. The groups were also given design toolkits consisting of a variety of modelling tools such as roundabouts, pedestrian crossings, landscaping material, modelling clay, and marking pens (see Figure 24.7). Included in each toolkit were also booklets that illustrated a wide range of road designs and their effects on traffic. The groups were given 45 minutes to work on their designs. At the end of that time each group was asked to nominate a spokesperson to describe their designs. After each group presented their designs the audience (residents who observed the workshop without participating in the designs) were invited to provide feedback and vote for a 'winning' design (votes were not counted and both teams were declared winners). The workshop activities were video recorded for later use.

Design Concepts

Based on the ideas produced in the participatory design workshops, we developed a series of conceptual designs to address the issues raised. The designs included a range of threshold and shared space treatments intended to slow traffic on the arterial road at the main shopping area: delineation of pedestrian walkways and cycle lanes to safely connect residential areas, recreational areas, and schools; and redesigned school bus stops. The conceptual designs had a common theme of delineating safe corridors for different types of road users including pedestrians, cyclists, local vehicular traffic, and through traffic: a suburban transport hierarchy. An example of some of the design alternatives are shown in Figure 24.8. The concept designs were then displayed in the library of the local school and parents and residents were invited to add further comments and suggestions. At this stage of the project a reorganization at the district council and ensuing budget cutbacks meant that further work on the project had to be postponed for the foreseeable future. The project designs and processes leading to the designs were documented

Figure 24.7   Participatory design toolkit (left) and community workshops in progress (centre and right)

to preserve the opportunity of resurrecting the project in the future when and if the financial situation changes.

Figure 24.8   Concept design alternatives produced from workshop results. Shopping area treatments (left) and school bus stop treatments (right)

Conclusion

The goal of these projects was to develop and demonstrate an SER approach that would establish a clear multi-level road hierarchy, with each level possessing a distinct 'look and feel' and discriminably different speed profiles. By these criteria, the projects were successful. Only one of the projects was carried through to completion, but for that area the effect of the road designs on vehicle speeds was to afford distinct speed profiles for each road category. At a subjective level, local roads now feel very different than before, in part because traffic speeds are much lower and in part because the landscaping and community islands (containing a variety of local art) provide a more residential atmosphere. Also of interest was whether the process could produce cost-effective designs to create distinct speed profiles for the collector and local roads. In this regard the project was successful as well; the cost of the SER treatments was lower than that required to install localized road humps on the local roads. Although the second project was carried through only to the production of design concepts, the involvement of the local residents to identify transport issues, create design alternatives, and ultimately delineate a suburban transport hierarchy for the area can also be considered a success. The participatory design techniques developed for this project have subsequently been applied to several other road safety upgrade projects. Finally, for both projects, the SER treatments made maximum use of residents' local knowledge and endemic road features intended to reflect and reinforce the character of the area rather than introduce site-specific interventions or unfamiliar road designs that would appear out of context with the area.

Acknowledgements

This research was funded in part by the Foundation for Research Science and Technology as project TER 0701. Projects such as these obviously require the work of a large number of people. I would particularly like to acknowledge the efforts of my colleagues at TERNZ Ltd., Hamish Mackie and Peter Baas; at Auckland City Council, Karen Hayes, Miguel Menezes, and Claire Dixon; at Rodney District Council, Gareth Hughes, Sarah Burrows, Shane Dale, and Jennifer Esterman; as well as Alan Dixon of the Ministry of Transport and Bill Greenwood of the New Zealand Transport Agency. I would also like to thank the many residents and community leaders who contributed to this project; it would not have been possible without their invaluable insights and assistance.

References

Aarts, L.T., and Davidse, R.J. (2007). Behavioural effects of predictable rural road layout: A driving simulator study. In *Proceedings of the International Conference Road Safety and Simulation RSS2007*. Rome: CRISS.

Brandt, E. (2006). Designing exploratory design games: A framework for participation in participatory design? In *Proceedings of the Ninth Participatory Design Conference 2006*. Trento, Italy: ACM.

Charlton, S.G. (2004). Perceptual and attentional effects on drivers' speed selection at curves. *Accident Analysis and Prevention, 36*, 877–84.

Charlton, S.G. (2007). Delineation effects in overtaking lane design. *Transportation Research Part F, 10*, 153–63.

Charlton, S.G., Mackie, H.W., Baas, P.H., Hay, K., Menezes, M., and Dixon, C. (2010). Using endemic road features to create self-explaining roads and reduce vehicle speeds. *Accident Analysis and Prevention, 42*, 1989–98.

Elliott, M.A., McColl, V.A., and Kennedy, J.V. (2003). *Road Design Measures to Reduce Drivers' Speed via 'Psychological' Processes: A Literature Review*. TRL Report 564. Crowthorne: Transport Research Laboratory.

Floyd, C., Mehl, W-M., Reisin, F-M., Schmidt, G., and Wolf, G. (1989). Out of Scandinavia: Alternative approaches to software design and system development. *Human–Computer Interaction, 4*, 253–350.

Goldenbeld, C., and van Schagen, I. (2007). The credibility of speed limits on 80 km/h rural roads: The effects of road and person(ality) characteristics. *Accident Analysis and Prevention, 39*, 1121–30.

Hardie, G.J. (1988). Community participation based on three-dimensional simulation models. *Design Studies, 9*, 56–61.

Herrstedt, L. (2006). Self-explaining and forgiving Roads: Speed management in rural areas. In *Proceedings of the 22nd ARRB Conference.* Canberra: ARRB Group Ltd.

Kaptein, N.A., and Claessens, M. (1998). *Effects of Cognitive Road Classification on Driving Behaviour: A Driving Simulator Study.* (MASTER. Deliverable 5.) Finland: VTT.

Lewis Evans, B., and Charlton, S.G. (2006). Explicit and implicit processes in behavioural adaptation to road width. *Accident Analysis and Prevention, 38*, 610–17.

Maarttola, I., and Saariluoma, P. (2002). Error risks and contradictory decision desires in urban planning. *Design Studies, 23*, 455–72.

Macbeth, A.G. (2007). A national road hierarchy: Are we ready? In *Proceedings of the IPENZ. Transportation Conference 2007.* Auckland, NZ: Transportation Group of the Institution of Professional Engineers of New Zealand.

Reich, Y., Konda, S.L., Monarch, I.A., Levy, S.N., and Subrahmanian, E. (1996). Varieties and issues of participation and design. *Design Studies, 17*, 165–80.

Rothengatter, J.A. (1999). Road users and road design. In G.B. Grayson (Ed.), *Behavioural Research in Road Safety IX* (pp. 86–93). Crowthorne: Transport Research Laboratory.

Tang, K.X., and Waters, N.M. (2005). The Internet, GIS and public participation in transportation planning. *Progress in Planning, 64*, 7–62.

Theeuwes, J. (1998). Self-Explaining Roads: Subjective categorization of road environments. In A. Gale (Ed.) *Vision in Vehicles VI*, (pp. 279–88). Amsterdam: North Holland.

Theeuwes, J., and Godthelp, H. (1995). Self-explaining roads. *Safety Science, 19*, 217–25.

Wegman, F., Dijkstra, A., Schermers, G., and van Vliet, P. (2005). *Sustainable Safety in the Netherlands: The Vision, the Implementation and the Safety Effects.* (Report R-2005-5.) Leidschendam, the Netherlands: SWOV Institute for Road Safety Research.

Weller, G., Schlag, B., Friedel, T., and Rammin, C. (2008). Behaviourally relevant road categorization: A step towards self-explaining rural roads. *Accident Analysis and Prevention, 40*, 1581–88.

Chapter 25

Behavioural Adaptation as a Consequence of Extended Use of Low-Speed Backing Aids

Christina M. Rudin-Brown
Ergonomics and Crash Avoidance Division, Road Safety and Motor Vehicle Regulation, Transport Canada, Ottawa, Canada and Human Factors Team, Monash University Accident Research Centre, Melbourne, Australia

Peter C. Burns
Ergonomics and Crash Avoidance Division, Road Safety and Motor Vehicle Regulation, Transport Canada, Ottawa, Canada

Lisa Hagen, Shelley Roberts and Andrea Scipione
CAE Professional Services, Ottawa, Canada

Introduction

Although they allow good forward visibility from the driver's seat, larger vehicles such as sport-utility vehicles (SUVs) and mini-vans tend to limit visibility to their rear and sides. Further, contemporary designs of all vehicle types, intended to improve structural integrity, aesthetics and aerodynamics, and enhance vehicle crashworthiness, may have inadvertently resulted in restricted visibility from the driver's seat (Quigley et al., 2001). These newer vehicles, combined with the ageing of the driving population, could conceivably result in an increase in the frequency and severity of backover crashes. However, because many of these crashes occur on private property and, consequently, may not be reported in official collision databases, the magnitude of the safety problem can only be roughly estimated. In the United States, there are approximately 228 fatalities and 17,000 injuries per year resulting from backover crashes involving light vehicles (NHTSA, 2010). In December 2010, the US National Highway Traffic Safety Administration (NHTSA) published a Notice of Proposed Rulemaking whereby it tentatively concluded that 'providing the driver with additional visual information about what is directly behind the driver's vehicle is the only effective near-term solution at this time to reduce the number of fatalities and injuries associated with backover crashes' (NHTSA, 2010: 76187).

Backing aids (also known as parking aids) refer to sensor or video-based in-vehicle systems that are intended to assist drivers in performing low-speed backing manoeuvres. Sensor-based backing aids use radar or ultrasonic sensors and provide some form (usually auditory) of warning to communicate the presence of, and distance to, obstacles located behind a vehicle (Green and Deering, 2006). These systems have the ability to reduce low-speed collisions between vehicles and objects and pedestrians so long as they detect them reliably and are used appropriately by drivers; however, because of limitations in sensor capabilities, they are not capable of reliably detecting pedestrians, and are thus intentionally marketed as convenience, rather than safety, systems (Glazduri, 2005). Most drivers are unaware of these limitations.

Visual-based backing aids use a video camera mounted in the rear bumper area that projects to a small monitor mounted in the instrument cluster or within the rear view mirror. Interestingly, rear view video systems are noted by NHTSA as 'the most effective technology option' to expand the driver's rear field of view to enable the driver to detect the area behind the vehicle (NHTSA, 2010). Rear view cameras may allow drivers to detect unexpected and unseen obstacles while backing; however, caution must be used when estimating safety benefits as these systems require direct glances to an in-vehicle display that is often located outside a driver's line of sight. Regardless of backing aid type, system warnings and information require attention and an appropriate response in a timely manner to be effective (Mazzae and Garrott, 2007).

The effectiveness of backing aids has been evaluated experimentally. In a recent study, drivers using one of five types of backing aid in a real world setting underwent a 'ruse' in which a staged obstacle (two-dimensional life-size image of a small child) was surreptitiously triggered to appear behind a test vehicle. Rear view video systems with the screen located in the rear view mirror were found to be associated with the largest reduction in crashes (Mazzae, 2010). Using a similar methodology with drivers of mini-vans, a rear view video system was associated with a 28 per cent reduction in crashes with an unexpected obstacle compared to using no system (Mazzae et al., 2008).

When used in the context of transportation psychology, 'behavioural adaptation' (BA) refers to changes in behaviour that occur after a particular change in the road traffic system. While always perceived as beneficial in some respect to an individual driver, it is the collective negative consequences of BA that are typically of primary concern to road safety professionals (Brown and Noy, 2004; Rudin-Brown and Noy, 2002; OECD 1990). Although backing aids are designed to assist drivers, and to generally improve road safety, their use may produce BA. For example, in a telephone survey, 11 per cent of owners of vehicles equipped with sensor-based parking aids reported believing that the system might increase the likelihood of a collision by causing driver over-reliance (Llaneras, 2006). These same drivers also reported fewer glances to their rear view and side mirrors, and fewer direct glances, while backing. Another survey (Jenness et al., 2007) revealed that 12 per cent of owners admitted that they had over-relied on their backing aid (by using just the system without checking their mirrors or turning to look out the rear window) within the previous two-week period. Objective measures of drivers' over-reliance confirm these findings. Drivers using sensor-based devices made fewer glances to their rear view and side mirrors, and conducted fewer direct glances while backing, than when unassisted (Lerner et al., 2007). Potential forms of BA to backing aids would likely be related to a driver's trust in the system, which is directly influenced by the reliability of a system's warnings. Research has shown that drivers may not always respond appropriately to warnings provided by in-vehicle collision warning systems, and a system's perceived reliability can drastically affect driver acceptance and trust in the technology (Rudin-Brown and Noy, 2002; Rudin-Brown and Parker, 2004). With backing aids, many drivers will ignore warnings when they come from systems that previously sounded false, or nuisance alarms (Llaneras et al., 2005), suggesting that a history of false warnings may degrade drivers' ability to respond by increasing their reaction time or by causing them to not respond at all. Further, many drivers will ignore a warning from a backing aid if s/he does not expect to encounter an object in their path (McLaughlin et al., 2003; Llaneras, 2006).

The main objective of the present study was to investigate whether behavioural adaptation, in terms of reduced visual search behaviour, would occur in drivers following extended use of sensor- and visual-based backing aids. It was hypothesised that, as a result of the reduction in effort needed to perform the backing manoeuvres when using the aids, drivers would demonstrate BA in terms of reduced visual search behaviour. It was also hypothesised that this reduction in visual search behaviour would be associated with collisions with an unexpected obstacle in conditions where the backing aids were active.

Method

Participants

Forty-two parent-aged (25–50 years; 19 female) drivers drove their own vehicle for a period of two months, throughout which it was equipped with one of three types of commercially available, aftermarket backing aid: ultrasonic ('sonar'; 15 participants), dash-mounted video ('video'; 15 participants), or rear view mirror-mounted video ('rear view'; 12 participants).

Procedure

Before and after the eight-week study period, driving performance was observed while performing a series of backing manoeuvres (parallel parking, perpendicular parking, and extended backing), both with and without the use of the backing aid (Figure 25.1). At the ostensible conclusion of testing, an 'unexpected obstacle' was placed behind the participant's vehicle without their knowledge, and they were asked to reverse out of the parking spot with their backing aid activated. All test sessions were conducted in Ottawa, Canada in a vacant, covered parking garage between January and June, 2009. Dependent measures included parking accuracy (distance to boxes and number of boxes hit) and visual search behaviour (glances to rear view mirror, glances to side mirrors, glances to front, side, and rear of vehicle, and shoulder checks) during backing manoeuvres. Throughout the test period, drivers were also required to complete a daily log book regarding their subjective experiences with using the backing aid.

Figure 25.1   Example of study set up (extended backing manoeuvre)

Statistical Analyses

Mixed analyses of variance (ANOVAs) were carried out to explore the influence of backing aid type (three levels; between-subjects), cue height (two levels; within-subjects), backing manoeuvre (three levels; within subjects), and session (two levels; within-subjects) on parking performance and visual search data. For the purposes of the present chapter, only those findings relating to experience with the backing aid (session) are reported. Non-parametric (Chi-square, Mann-Whitney) analyses were conducted on the frequency of crashes with the unexpected obstacle, and the logbook subjective data. In all cases, a two-tailed alpha-level of .05 was used to determine statistical significance. Comparisons between conditions were made using 95 per cent confidence intervals (Masson and Loftus, 2003; Jarmasz and Holland, 2009).

Results

Parking Accuracy

To assess changes in parking accuracy due to the extended use of a backing aid, data collected during the pre-use session with backing aid turned off were compared to data from the post-use session with the backing aid on. The main effect of the session was significant for distance to boxes, $F(1,39) = 15.1, p < .01$, and number of boxes hit during the manoeuvres, $F(1,39) = 15.9, p < .001$ (Figure 25.2). Regardless of backing aid type, participants parked their vehicles closer to the boxes, and hit fewer boxes, in the post-use session with the backing aid turned on compared to the pre-use session with backing aid off.

Visual Search Behaviour

To assess changes in visual search behaviour due to the extended use of a backing aid, glance data collected during the pre-use session with the backing aid off were compared to data from the post-use session with the backing aid on. The main effect of session was significant for the total number of glances, $F(1,38) = 34$, $p < .001$, with participants engaging in significantly less visual search behaviour during the post-use session with the backing aid on compared to the pre-use session with backing aid off (Figure 25.3).

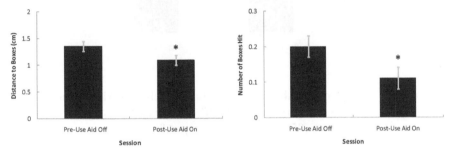

Figure 25.2 **Effect of backing aid on parking accuracy: distance to boxes (left); number of boxes hit (right) (95% CIs shown; * = p <.01)**

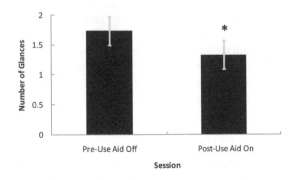

Figure 25.3 **Effect of backing aid on visual search behaviour: number of glances for 'pre-use / aid off' compared to 'post-use / aid on' conditions (95% CIs shown; * = p <.001)**

To assess whether the observed reduction in visual search behaviour, following the two-month study period, continued despite the backing aid being turned off, glance data collected during the pre-use session with backing aid off were compared to glance data from the post-use session with backing aid off. The main effect of session was significant for the total number of glances, $F(1,39) = 34$, $p < .001$, with participants engaging in significantly less visual search behaviour in the post-use session with backing aid off compared to the pre-use session with backing aid off (Figure 25.4).

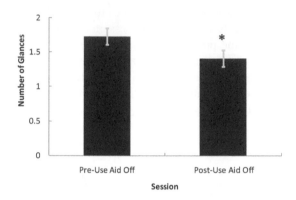

Figure 25.4   Effect of backing aid on visual search behaviour: number of glances for 'pre-use / aid off' compared to 'post-use / aid off' conditions (95% CIs shown; * = p <.001)

Reaction to Unexpected Obstacle

At the ostensible conclusion of testing, but before the backing aid was removed from the vehicle, an 'unexpected obstacle' was placed behind the participant's vehicle without their knowledge, and they were asked to reverse out of the parking spot with their backing aid activated. The majority of participants failed to detect the unexpected obstacle, resulting in a collision. There were no significant differences among the groups in terms of the percentage of participants who collided with the object, $\chi^2 (2) = 1.48, p > .05$ (Figure 25.5).

Subjective Data

Throughout the study period, participants were instructed to record their daily driving activities and perceptions of the backing aids in log books. While there were no significant differences among the three backing aid groups in terms of reported backing behaviour (meaning the number and type of backing episodes undertaken per day), there were several findings of interest. Participants in the

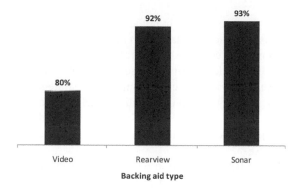

Figure 25.5 Percentage of participants who collided with the unexpected object

sonar group reported using the backing aid significantly more often throughout the study period (mean = 188 times) than participants in either the rear view (67 times) or the video (100 times) groups, F (2,36) = 3.36, p < .05. In addition, participants who reported turning off their backing aid at some point during the trial were significantly less likely, than those who did not report turning off their backing aid, to hit the unexpected obstacle in the post-test (U = 35, n$_1$ = 14 n$_2$ = 24, p < .05). Finally, participants who reported turning off their backing aid were also more likely, than those who did not report turning off their aid, to report experiencing frustration with their backing aid during the study period (U = 103, n$_1$ = 33 n$_2$ = 5, p = .05).

Discussion

As expected, extended use of backing aids resulted in more accurate parking. Participants parked closer to, and collided with fewer, boxes when the backing aids were active in the post-test session than when they drove unassisted during the pre-test session. Although parking accuracy improved, following the eight-week period drivers engaged in significantly less visual search behaviour while backing when they used the devices than when performing the same manoeuvres unassisted at the beginning of the study. This difference persisted even when the backing aids were turned off, suggesting that experience with the backing aids resulted in drivers adapting their visual search behaviour in a potentially safety critical manner. Further, regardless of backing aid type, the majority (80–93 per cent) of participants collided with the unexpected obstacle at the end of the post-test session.

The visual search results, in particular the finding that the number of glances did not return to baseline levels in the post-test session even when the aids were turned off, provide support for the hypothesis that drivers demonstrate BA to backing aids in terms of reduced visual search behaviour. Caution should therefore be used when attempting to estimate the overall safety benefits of backing aids. It is possible that, with extended real-world use, drivers may learn to over-rely on the backing aids resulting in less effective visual search strategies. In the event that a backing aid that is used by such a driver fails to detect an object or pedestrian, or when an object is not expected (such as in the unexpected obstacle condition in the present study), it is likely that the driver will fail to perceive it directly, and crash risk will increase.

The possibility that BA developed in response to the extended use of backing aids, in terms of reduced visual search behaviour and an associated increase in crash risk, is also supported by the unexpected obstacle event data, where the majority of participants collided with the object despite having the backing aid turned on. The large number of collisions observed in this scenario may have been due to a delay in system activation (for the video-based systems) or because of the unexpected nature of the event, which caused drivers to ignore warnings from the ultrasonic device. Whatever the reason for the response, results are consistent with previous backing aid research (Mazzae, 2010; Mazzae et al., 2008; Llaneras et al., 2005; McLaughlin et al., 2003), and provide further support for the contention that backing aid marketing efforts should remain focused on their convenience, rather than their safety, features.

Finally, subjective data collected from the driver log books provide some indication that a possible mechanism underlying the observed BA is driver trust in the system. Drivers who reported turning off their backing aid during the study period were less likely to collide with the unexpected obstacle than were drivers who did not turn off their aid. Also, participants who reported turning off their backing aid were more likely than those who did not, to report experiencing frustration with their backing aid during the study period. Both of these findings, albeit suggestive, can be interpreted as surrogate measures of drivers' trust in, and consequent dependency on, backing aids. Logbook comments for all groups indicated that participants reported not relying on the backing aids and mistrusting the feedback provided by the aids. Further, due to numerous false alarms, participants in the sonar group reported having very little confidence in that aid's ability to accurately detect objects behind their vehicles. Despite these comments, however, participants in all groups trusted the systems enough to rely on them over their own vision by engaging in more limited visual search behaviour when the backing aids were used compared to when they were not active.

A limitation of the study that should be noted is related to the unreliability and limited usability of the aftermarket backing aids used. Although the three models were selected to represent a variety of commercially available devices, participants in all groups reported issues relating to their reliability and, in the case of the visual-based systems, the clarity of the display. Nevertheless, the use of a within-subjects study design with participants acting as their own controls limited the influence of these factors on the findings. Future research should look at driver performance when using factory-installed, original equipment, as well as the next generation of backing aids, which not only displays video, but warns the driver using image processing capabilities. Naturalistic driving study designs would also be beneficial.

Collectively, results indicate that drivers exhibit BA in terms of reduced visual search behaviour following the extended (two-month) use of the three aftermarket backing aids used in this study. While results show that this extended use does improve parking accuracy, backover crash risk associated with the extended use was very high, with between 80 and 93 per cent of participants colliding with an unexpected obstacle, despite the presence of an active backing aid. Consumers need to be adequately informed of the limitations of these devices, and the potential for BA to develop as a consequence of their use, in order to limit the likelihood of real world backover crashes. Further, visual-based backing aids should use cameras that provide an adequate rear field-of-view of the rear of the vehicle and beyond, include large enough display screens with clear resolution in all lighting conditions, and have fast system start-up times. Finally, drivers are reminded that backing aids are not a substitute for parental supervision or good driving practice; drivers of all vehicles, including those equipped with backing aids, should always check behind their vehicles before they get in.

Acknowledgements

The authors would like to thank Ms Cory Tam and Ms Erica Elderhorst for assisting with the analysis of the driver log books. The views presented in this chapter are those of the authors and not necessarily of Transport Canada.

References

Brown, C.M., and Noy, Y.I. (2004). Behavioural adaptation to in-vehicle safety measures: Past ideas and future directions. In: T. Rothengatter and R.D. Huguenin (Eds.), *Traffic and Transport Psychology: Theory and Application*, pp. 25–46, Netherlands: Elsevier.

Glazduri, V. (2005). An investigation of the potential safety benefits of vehicle backup proximity sensors. *Proceedings of the 19th International Technical Conference on the Enhanced Safety of Vehicles*, Washington, DC, pp. 1–10.

Green, C.A., and Deering, R.K. (2006). *Driver performance research regarding systems for use while backing.* Society of Automotive Engineers Paper No. 2006–01–1982. Warrendale, PA.

Jenness, J.W., Lerner, N.D., Mazor, S., Osberg, J.S., and Tefft, B.C. (2007). *Use of advanced in-vehicle technology by young and older early adopters: Survey results on sensor-based backing aid systems and rear view video cameras.* NHTSA Report No. DOT HS 810 828. Washington, DC.

Jarmasz, J., and Hollands, J.G. (2009). Confidence intervals in repeated measures designs: The number of observations principle. *Canadian Journal of Experimental Psychology, 63*, 124–38.

Lerner, N.D., Harpster, J.L., Huey, R.W., and Steinberg, G.V. (2007). Driver backing-behavior research: Implications for backup warning devices. *Transportation Research Record, 1573*, 23–9.

Llaneras, R.E., Green, C.A., Kiefer, R.J., Chundrlik, W.J., Altan, O.D., and Singer, J.P. (2005). Design and evaluation of a prototype rear obstacle detection and driver warning system. *Human Factors, 47*, 199–215.

Llaneras, R.E. (2006). *Exploratory study of early adopter, safety-related driving with advanced technologies*. NHTSA report No. DOT HS 809 972. Washington, DC.

McLaughlin, S.B., Hankey, J.M., Green, C.A., and Kiefer, R.J. (2003). Driver performance evaluation of two rear parking aids. *Proceedings of the 18th International Technical Conference on the Enhanced Safety of Vehicles*, Nagoya, Japan, pp. 1–9.

Masson, M.E.J., and Loftus, G.R. (2003). Using confidence intervals for graphically based data interpretation. *Canadian Journal of Experimental Psychology, 57*, 203–20.

Mazzae, E.N. (2010). *Drivers' use of rearview video and sensor-based backing aid systems in a non-laboratory setting*. NHTSA research report in docket number NHTSA-2010–0162, December.

Mazzae, E.N., Barickman, F., Baldwin, G.H.S., and Ranney, T. (2008). *On-road study of drivers' use of rearview video systems (ORSDURVS)*. NHTSA report No. DOT HS 811 024, Washington, DC, September.

Mazzae, E.N., and Garrott, W.R. (2007). *Experimental evaluation of the performance of available backover prevention technolgies for medium straight trucks*. NHTSA report No. DOT HS 810 865. Washington, DC.

NHTSA (2010). Docket number NHTSA-2010–0162. *Federal Register, 75(234)*, 76186–76250, December 7.

OECD (Organization for Economic Cooperation and Development). (1990). *Behavioural Adaptations to Changes in the Road Transport System*. Paris: OECD Road Research Group.

Quigley, C., Cook, S., and Tait, R. (2001). *Field of vision (A-pillar geometry): A review of the needs of drivers*. UK Department of the Environment, Transport and the Regions Report No. PPAD 9/33/39, January.

Rudin-Brown, C.M., and Parker, H.A. (2004). Behavioural adaptation to adaptive cruise control (ACC): Implications for preventive strategies. *Transportation Research Part F, 7*, 59–76.

Rudin-Brown, C.M., and Noy, Y.I. (2002). Investigation of behavioral adaptation to lane departure warnings. *Transportation Research Record, 1803*, 30–37.

Chapter 26

Enhancing Sustainability of Electric Vehicles: A Field Study Approach to Understanding User Acceptance and Behaviour

Thomas Franke, Franziska Bühler, Peter Cocron,
Isabel Neumann and Josef F. Krems
Cognitive and Engineering Psychology,
Chemnitz University of Technology, Germany

Introduction

Increasing concern about the environmental impact of current road transport systems as well as the analyses of risks associated with peak oil (Hirsch et al., 2005) have led to greater interest in sustainable transportation. Mobility systems based on electric vehicles (EVs) are considered promising for coping with these challenges. Many countries have set up action plans to increase the proportion of EVs within the road transport sector (Die Bundesregierung, 2009). However, the positive effects of EVs on the sustainability of the transport sector are still debated (Horst et al., 2009; Huo et al., 2010). It has been shown that these effects are largely dependent on how an electric mobility system (EMS) is set up as well as how it is used (Eggers and Eggers, 2011). Hence the user is a critical parameter in the equation of the net environmental benefit of EMSs. Adopting the perspective of human-centred systems engineering (Nemeth, 2003) three essential components should to be taken into account when optimizing user-system interaction: technical system, user and task. The task of EMSs is to increase sustainability. Much is known about the theoretical sustainable potential of certain technical system variants for sustainability, but little is known about user factors in this equation.

The objective of the large-scale field study outlined in this chapter was to gain a comprehensive understanding of the user-driven dynamics of sustainability in EMSs. Acceptance of favourable system layouts and efficient interactions with system resources are key dimensions. Adaptation processes occurring on these dimensions as well as personal characteristics (for example dispositions, traits, needs) may yield promising variables to increase our understanding of the dynamics in these dimensions related to sustainability.

First, important success factors and associated research topics which have evolved from these factors will be structured. Then, the field study approach is outlined with a detailed account of study methodology. Thereafter, focused analyses on selected research topics follow. A final discussion and outlook concludes the chapter.

Factoring Users into the Equation of EMS Sustainability

Sustainability is a multidimensional concept. Three pillars should be addressed: environmental protection, social development and economic development (UN General Assembly, 2005). The sustainability of EVs has been intensely discussed, particularly with respect to the environmental dimension. Lower local noise and exhaust emissions are the two least contested factors. However, significant reductions in greenhouse gas emissions depend on the energy source used for propulsion and production (Holdway et al., 2010). Although there are contradictory findings in the literature (Brady and O'Mahony, 2011; Thiel et al., 2010) there is a tendency to believe that the environmental benefits of EMSs will be negligible for a fossil-fuel-based scenario (Huo et al., 2010). However, even under such a worst-case scenario EVs may still have some utility on social and economic dimensions because of their potential to mitigate risks associated with peak oil prices (Hirsch et al., 2005). In addition, EMSs can help to integrate renewable energy into the grid by acting as flexible energy storage buffers and thereby promoting sustainability of the general energy supply.

Hence, there are considerable degrees of freedom concerning the sustainability that EVs offer. We argue that the (potential) user is a critical factor in this equation. As a best-case scenario, we assume that the maximum benefit for sustainability would be achieved if all car buyers with high daily EV-capable mileage would switch to an EMS with a sustainable layout. Such a setup would mean low resources and emissions needed for production (for instance battery size determined according to actual user need) and operation (such as exclusive use of excess energy from renewable sources). The mobility resources that such an EMS could offer would then have to be utilized in an energy efficient manner. This could be accomplished if all individual mobility needs were transferred to this EMS, with optimal utilization of range and recharging options. Moreover, this successful adaptation would require ecological behaviour while using the system (that are eco-driving, treating battery for longevity, assisting excess energy use by exhibiting appropriate mobility and charging patterns). Interacting in this way with the EMS should not compromise the quality of the user experience and should ensure acceptance so that users still act as disseminators. Ideally such setups should be amenable to multi-modal and multi-person usage scenarios (for example intuitive design of user interfaces in whole EMSs).

We propose that two factors should be distinguished in interrogating and optimizing the user as a factor in the sustainability of EMSs: (1) Users should prefer sustainable system layouts, and (2) users should efficiently interact with system resources to assure that the sustainable potential of a given setup is maximized. Assumptions of the best-case scenario described earlier together with these two proposed factors resulted in the following research questions.

First, concerning the acceptance (positive attitudes and purchase intention) of sustainable EMS layouts, research questions are: Is there a general high acceptance concerning an essentially sustainable EMS? Do users accept and actively prefer the use of excess energy from renewable sources? Do they accept lower buffers in terms of mobility resources (range and recharging options)? Do users prefer low noise-setups or do they experience a conflict between low noise-emissions and traffic safety?

Second, research questions that address the utilization of the sustainable potential of a given setup, by optimal user-system interactions include: Do users transfer significant shares of mobility to EMSs? Do drivers optimally utilize available mobility resources (for example, range, recharging options)? Do users consider longevity in their treatment of the EMS?

Variance within and between these variables may be explained by several moderator variables, which in turn can inform strategies to optimize sustainability of EMSs. Research questions concerning these moderating effects are: Does experience (such as practice, knowledge) moderate score values? Do personal variables (such as traits, dispositions, needs) yield interaction effects? Is suboptimal behaviour more a function of conflicting behavioural intentions or of behavioural abilities? Insights provided by these questions can help to quantify the real-world impact of the user factor on EMS sustainability and to identify promising variables for its enhancement.

A Field Study Approach to Understanding User-driven Dynamics in Sustainability of EMS

Research on factors of EMS sustainability previously focused on modelling the user factors with statistical data, such as data from travel surveys (Brady and O'Mahony, 2011). In addition, studies of potential car-buyers with little knowledge of EMSs have dealt with acceptance-related topics (Chéron and Zins, 1997), with some studies simulating experience with reflexive methodologies (Kurani et al., 1996). However, these studies are limited in predicting what car-buyers in markets with already experienced customers (after societal adaptation to EMS has occurred) will accept and in quantifying the dynamics introduced by user behaviour. As a result field studies have been requested which yield adapted interaction patterns with experienced users in realistic settings (Kurani et al., 1994). These studies also have methodological limitations (Golob and Gould, 1998), but are nevertheless accepted for investigating the user perspectives on EMSs, while also providing validity for acceptance related issues in markets that do not exclusively comprise inexperienced first-time buyers.

The methods outlined in the following refer to a large-scale EMS field study incorporating two subsequent user studies which took place in the Berlin metropolitan area in 2009 and 2010 (see also Cocron et al., forthcoming). Addressing a major criticism of EMS field studies, several measures were taken to ensure high ecological validity for EMS use in the coming years. An electric vehicle was paired with private and public charging infrastructure supporting the use of excess energy from renewable sources in a metropolitan area. Users were recruited via an online screener application that was publicized via advertisements in newsprint and online media. Subjects were selected according to several must-have criteria and further distribution criteria that aimed to prevent restriction of variance on several basic sociodemographic and mobility-related variables. Users had to agree to pay a monthly leasing fee over the six-month study period. Study length was designed to allow for full adaption to the EMS. Hence the resulting sample is expected to reflect the viewpoint of early customers of EMSs.

In addition a multi-method approach that incorporated structured qualitative interviews, questionnaires, diary methods, experimentally oriented methods and continuous data logging of vehicles and charging infrastructure values offered the possibility of data triangulation and fusion, filling gaps in scientific knowledge on user experience and behaviour in EMSs. For these analyses a main-user approach was adopted such that in a household only main EV user data was collected whereby a special car key was assigned for subsequent allocation of data logger data.

Sequence of Data Collection and Method Corpus

Table 26.1 depicts the sequence of data collection events for each user over the period of the study. Data were collected before receiving the EV (T0), after three months of experience (T1), and after six months at vehicle hand back (T2). The sequence of events was the same in the two user studies except that a second test drive was applied only in the second user study (S2) and a few of the questions from the pre-experience interview were shifted to a preceding telephone interview in the first user study (S1). Most essential methodological elements were the same for both user studies with the method corpus increasing from 873 items in S1 questionnaires to 1,380 items in S2 (plus nearly 200 hours of verbatim-transcribed interview data). The following description focuses on S2 methodology as it is most representative, while noting essential differences to S1. A structured matrix of central topics and sub-topics of questionnaires and interviews is depicted in Figure 26.1.

In the online-screener all applicants were asked to submit basic sociodemographic and mobility-related information. After selection of users from the applicant list (for details on the sample see Neumann et al., 2010) and prior to vehicle handover (T0), each of the 40 resulting main users filled out a seven-day baseline travel diary, which was structured according to established travel survey designs (Kunert and Follmer, 2005). The diary was aimed at assessing

Table 26.1   Sequence of data collection events within user studies

Timeline (months)	Data collection event	Acronym
-3	*Screening*	
-0.5	Baseline diary week	T0
0	Pre-experience questionnaire and interview	T0
0	Test drive	T0
0	Post-test drive questionnaire and interview	T0
0	*Vehicle handover*	
2.5	Diary week after experience	T1
3	Questionnaire and interview after experience	T1
3	Test drive*	T1
5.5	Final diary week	T2
6	Final questionnaire and interview	T2
6	*Vehicle handback*	

* Second user study only.

1. Impression

Safety   appraisal | critical situations

System competence/   previous knowledge | technical background
knowledge   knowledge

Regenerative   adaptation | acceptance[1] | usage and influence on driving task | trust
braking

Eco-driving   acceptance of eco driving[1] | knowledge | user behaviour | driving style

Range   adaptation | utilization | satisfaction & preferences | psychological range levels and buffer values | knowledge | concerns

Charging   controlled charging (acceptance, behaviour) | handling | user-battery interaction | future concepts | battery lifetime | motives of charging | public charging

Acceptance   purchase intention | attitude towards EVs | usefulness & satisfaction[1] | influencing factors (attitude and prejudice of social network, joy of use, barriers)

Mobility   (mal)adaptation | mobility resources needs | satisfaction | future mobility concepts | mobility patterns

Personality   car-affinity | personal driving style[2] | control beliefs in dealing with technology[3] | sensation seeking[4] | affinity for technology[5] | uncertainty avoidance[6]

Acoustics   adaptation | safety evaluation | estimation of speed | critical situations | sound design

Human–machine   adaptation | usage | perceived usability | preferences | users'
interface/ handling   conceptions

Environmental value/   environmental concerns | renewable
renewable energies   energies (acceptance & preferences)

Demographics

[1] Van Der Laan et al., 1997 | [2] French et al., 1993 | [3] Beier, 1999 | [4] Beauducel et al., 2003 | [5] Karrer et al., 2009 | [6] Dalbert, 1999

Figure 26.1   Matrix of research topics examined in the field study. Only exemplary sub-topics are depicted

mobility patterns and determining indicators for mobility needs. This diary was applied again at T1 and T2 in an extended form designed for tracking changes in mobility patterns associated with the introduction of the EV to the main-user's mobility resources. In addition, the seven-day charging diary at T1 and T2 was designed to assess time, location, energy status and motivation related to each charging process. Much of the questionnaire content was also included in the semi-structured face-to-face interviews to assure high data quality. A standard six-point Likert scale (ranging from 1 = *Completely disagree* to 6 = *Completely agree*) was applied in scales constructed for the study. The first test drive was conducted to gain a qualitative picture of first impressions while driving the EV and to assess certain interaction skills, which were again assessed in the T1 test drive. Additional methods employed during interviews included, for example, a range game (Franke et al., forthcoming) and a choice-based conjoint analysis (Krems et al., 2010).

Focus-analyses of User Factors Affecting Sustainability of EMSs

A subset of the presented research questions are addressed in the following. For selected topics, moderator effects will also be discussed. Because of limitations of space, results of the first user study are presented exclusively. The second user study, however, revealed a comparable pattern of results.

Do Users Accept and Prefer Sustainable EMS Layouts?

Several scales were constructed and adapted from the literature (Bühler et al., 2011) to assess facets of EMS acceptance. Here we focus on a seven-item attitude scale (Bühler et al., 2011) and a single-item indicator for determining purchase intention, which stated 'I would spend a third more for an electric vehicle like the MINI-E than for a comparable vehicle with a conventional engine'. Attitudes, which were already very positive at T0, increased with experience (T0: $M = 4.62$, $SD = .71$; T1: $M = 4.91$, $SD = .74$) yielding a significant ($p < .05$) effect (Bühler et al., 2011). In terms of purchase intentions 64 per cent 'agreed' on that item at T0 (dichotomization of six-point Likert scale). This value decreased to 51 per cent at T1. The relatively high score values are probably supported by selection bias. However, increasing acceptance with experience might indicate that attitudes can be positively influenced by hands-on experience. Further qualitative results from the study suggest that even a short test drive can enhance attitudes toward EVs. However, decreases in purchase intentions may indicate increased price sensitivity when users are past the honeymoon phase. This problem could be ameliorated if users were willing to accept lower range capabilities thereby decreasing the additional costs of EVs in comparison to combustion vehicles.

Users' preferences in terms of energy sources for charging the EV were not in conflict with principles of sustainability. From the scales that were assessed (Rögele et al., 2010) the three-item scale on attitudes towards usage of renewable energies in EVs indicated high acceptance (T0: $M = 4.79$, $SD = 1.09$; T1: $M = 4.79$, $SD = 0.88$) and users indicated strong preferences for renewable energy sources (Rögele et al., 2010).

To increase market potential and environmental benefit it would be favourable if users accepted lower buffers in terms of mobility resources. Past research has shown that inexperienced users have very high preferences concerning range and recharging options (Eggers and Eggers, 2011). We also found range preferences to be relatively high in the present experienced user sample (available range under daily conditions was evaluated as just acceptable at T1: $M = 156$ km, $SD = 81$ km, $Q_3 = 180$; range evaluated as sufficient at T1: $M = 227$ km, $SD = 124$ km, $Q_3 = 300$). Only 37 per cent of users judged range setups, which were equal to or smaller than those offered by their current EV, as sufficient. Even in users who stated they could use the EV for > 95 per cent of daily trips, this number could only be increased to 53 per cent if passenger and luggage space were not restricted. A similar number (60 per cent) was yielded for users who completely agreed that the current range was sufficient for daily use.

Do users prefer low-noise EMSs or do they experience a conflict between low noise emissions and traffic safety? Within the present study, few critical incidents were reported by users: 68 per cent had no incidents and 84 per cent had < 10 incidents within the six-month study period (Cocron et al., 2011). Moreover, only a small proportion of users (13 per cent) preferred active sound design. In addition, low noise was perceived as one of the main advantages and attractions of EVs and thus might be essential for general acceptance of EMSs (Cocron et al., 2011). However, it remains to be verified if the general public and public authorities share the same positive appraisal of low noise emissions in EMSs.

Do Users Exploit the Sustainable Potential of EMSs to the Fullest Extent?

A central factor in sustainability is that users transfer the maximum possible share of their mobility needs to more sustainable transport options. Modal split is used as an indicator in this area of research (Schafer, 1998). For the present analyses data from travel diaries of 30 users were analysed that fitted the necessary conditions for comparing their data to available statistical data from nationwide surveys (Ahrens, 2009). Taking the baseline-data from T0, the user sample had relatively high shares of individual motor car traffic (trips by foot: 9.0 per cent, bike: 8.3 per cent, public transport: 12.2 per cent, individual motor traffic: 70.5 per cent) compared to panel data for the population of Berlin (by foot: 28.6 per cent, bike: 12.6 per cent, public transport 26.5 per cent, individual motor traffic: 32.3 per cent [Ahrens, 2009]). After the introduction of the EV as a sustainable transport option this distribution changed substantially (trips by foot: 5.4 per cent, bike: 1.7 per cent, individual combustion-powered motor traffic: 21.3 per cent, public transport: 1.8 per cent, EV: 70 per cent). Although temporal shifts should be interpreted with caution, the general pattern indicates that the EV could substantially reduce the share of individual combustion-powered motor traffic in the user sample. However, a reduction in proportion of the other sustainable transport options might indicate that these are less frequently used as a result of the availability of the more convenient option the EV offers.

Closely related to this topic, users should be willing and able to optimally utilize the valuable range resources that the EMS has to offer. Results show that users reserve a substantial buffer in the range they are willing to utilize (Franke et al., forthcoming). Within the sample users were only comfortable with utilizing a mean average of 82 per cent ($SD = 11$ per cent, $Q_1 = 76$ per cent, $Q_3 = 90$ per cent) of the range that they achieved under daily conditions. Interestingly, there was a moderate relation between the tendency to actively test the range ('I deliberately tried to exhaust the range in order to see how far I could go with the MINI E' at T2) and the acceptance of higher utilization of range resources. Thus, helping the user explore the range might also promote the use of available range resources.

Finally, the low and variable lifetime of batteries may present a barrier to purchase intentions and compromise the efficiency of resources needed for the production of EMSs. Although it is very difficult to quantify the influences of optimal user behaviour it can be assumed that charging and discharging patterns have a significant impact on battery life. In the present sample, 61 per cent of users agreed to an item stating 'Throughout the EV trial I tried to handle the battery in a way that would prolong its life.' Seventy-six per cent of users agreed when the context was revised to 'if I had paid for the EV outright'. Subjective knowledge (assessed as the mean confidence in nine ratings on the effectiveness of certain behaviours to prolong battery life, such as 'avoid frequent charging at high charge levels') was related to actual treatment for longevity (median split, 47 per cent tried to prolong battery life in the low knowledge group, and 74 per cent in the high knowledge group) indicating that an increase in user knowledge, or rather problem awareness, could lead to better interaction with the lifetime resources of the battery.

General Discussion and Outlook

This chapter has emphasized the importance of including the user as a factor in determining sustainability of EMSs. A field study approach outlined together with study results aimed to provide an overview of exemplary user factors in a state-of-the-art EMS. From these results it can be concluded that there are substantial variations in the user factor: between the different facets important for EMS sustainability and within users. In general it was found that users prefer and accept sustainable system setups and exploit the sustainable potential of the EMS in an acceptable way. However, there still seems to be substantial potential for increasing the sustainability of EMSs given the variance in the measured variables and the indicated moderating effects. Helping users make full use of given mobility resources is an important topic here. Guided exploration of mobility needs and mobility resources could be a fruitful approach as the experience-related moderator effects suggest.

The analyses presented herein gave an exemplary overview of user-driven factors in EMS sustainability. However further analyses are necessary to more closely examine and replicate findings with data triangulation. Selection bias is also an important consideration in the present study. We are currently testing the transferability of core results in follow-up studies, in users without private charging options and in study settings that allow more freedom in recruiting users. Furthermore we are working on replicating important results in additional experimental settings, such that causal conclusions may be made.

The present field study revealed that state-of-the-art EMSs are already well suited to a substantial sample of users. During the history of the EV, positive findings in user acceptance and positive evaluations of suitability for daily use have often led to the conclusion that stakeholders should just wait for the very next improvement in vehicle technology because this next step would then finally lead to high market potential and hence high sustainability of investing resources in the development and marketing of EMSs. However, we are still far from the situation where EMSs perfectly mimic combustion-powered mobility systems. Still, especially from the indications of high acceptance and price-sensitivity we conclude that further technical improvements should be implemented with the aim of reducing prices rather than increasing performance, hence reducing the entrance threshold for attracting new users. That is, users need to be sensitized to the special characteristics offered by EMSs, discouraged from internal combustion engine vehicle comparisons, and encouraged to consider the resources offered by EMSs in meeting their personal needs. This might be a societal adaptation process that could be supported by giving more people the chance to experience EMSs in real-life.

The present study also has implications for future research. For example, the paradoxical disparity of range satisfaction and range preferences, which has been attributed to different methodological approaches in the past (Kurani et al., 1994), has also been found, albeit less intensely, in this sample of users. This finding needs further investigation, for example, focusing on possible trade off effects. Also, this study might help inspire the development of indicators for the sustainability of EMSs that integrate the user as an important factor. It is also our aim to design more reliable, valid and economic indicators in future studies (for example extending modal split to EMSs in fleet settings).

As a final remark, we would like to clarify the perspective on EMSs adopted in this chapter. We do not wish to imply that it is the user's fault if an EMS does not reach its intended sustainable potential. Nor do we argue that it is the *user* who should adapt to the *system*. We believe that a positive, satisfying and joyful user experience is key to ensuring EMS sustainability. And it is much more favourable to adapt the system to the user than to adapt the user to the system (Good et al., 1984). However to reach this aim, a comprehensive understanding of the user factors in EMSs is indispensable.

Acknowledgements

This study was funded by the German Federal Ministry for the Environment, Nature Conservation and Nuclear Safety. We thank our consortium partners Vattenfall Europe AG (Dr C.F. Eckhardt, F. Schuth) and the BMW Group (Dr A. Keinath, Dr M. Schwalm), which made this research possible. A special thanks to Kristin Lange for her support during preparation of the manuscript and Jens Nachtwei for critical review of this chapter.

References

Ahrens, G-A., Ließke, F., Wittwer, R., and Hubrich, S. (2009). Sonderauswertung zur Verkehrserhebung 'Mobilität in Städten- SrV 2008': Städtevergleich [Special evaluation of traffic investigation mobility in cities SrV 2008: Comparison of towns and cities]. Dresden University of Technology, Germany.

Beauducel, A., Strobel, A., and Brocke, B. (2003). Psychometrische Eigenschaften und Normen einer deutschsprachigen Fassung der Sensation Seeking-Skalen, Form V [Psychometric properties and norms of a German version of the sensation seeking scales, form V]. *Diagnostica, 49*(2), 61–72.

Beier, G. (1999). Kontrollüberzeugungen im Umgang mit Technik [Control beliefs in dealing with technology]. *Report Psychologie, 9*, 684–93.

Brady, J., and O'Mahony, M. (2011). Travel to work in Dublin: The potential impacts of electric vehicles on climate change and urban air quality. *Transportation Research Part D: Transport and Environment, 16*, 188–93.

Bühler, F., Franke, T., and Krems, J.F. (2011). Usage patterns of electric vehicles as a reliable indicator for acceptance? Findings from a German field study. *Paper accepted to appear in Proceedings 90th Annual Meeting of the Transportation Research Board, Washington, D.C.*

Chéron, E., and Zins, M. (1997). Electric vehicle purchasing intentions: The concern over battery charge duration. *Transportation Research Part A: Policy and Practice, 31*, 235–43.

Cocron, P., Bühler, F., Franke, T., Neumann, I., and Krems, J.F. (2011). The silence of electric vehicles: Blessing or curse? *Paper accepted to appear in Proceedings of the 90th Annual Meeting of the Transportation Research Board, Washington, D.C.*

Cocron, P., Bühler, F., Neumann, I., Franke, T., Krems, J.F., Schwalm, M., and Keinath, A. (forthcoming). Methods of evaluating electric vehicles from a user's perspective: The MINI E field trial in Berlin. *IET* Intelligent Transport Systems.

Dalbert, C. (1999). Die Ungewissheitstoleranzskala: Skaleneigenschaften und Validierungsbefunde [Tolerance of ambiguity scale: Scale characteristics and validity]. University of Halle, Germany.

Die Bundesregierung (2009). *The Federal Government's National Development Plan for Electric Mobility.* Published online in August 2009. [Online]. Available at: http://www.bmu.de/files/pdfs/allgemein/application/pdf/nep_09_bmu_en_bf.pdf, accessed 31 January 2011.

Eggers, F., and Eggers, F. (2011). Where have all the flowers gone? Forecasting green trends in the automobile industry with a choice-based conjoint adoption model. *Technological Forecasting and Social Change, 78,* 51–62.

Franke, T., Neumann, I., Bühler, F., Cocron, P., and Krems, J.F. (forthcoming). *Experiencing range in an electric vehicle: Understanding psychological barriers.* Chemnitz University of Technology, Germany.

French, D.J., West, R.J., Elander, J., and Wilding, J.M. (1993). Decision-making style, driving style and self-reported involvement in road traffic accidents. *Ergonomics, 36,* 627–44.

Golob, T.F., and Gould, J. (1998). Projecting use of electric vehicles from household vehicle trials. *Transportation Research Part B: Methodological, 32,* 441–54.

Good, M.D., Whiteside, J.A., Wixon, D.R., and Jones, S.J. (1984). Building a user-derived interface. *Communication of the ACM, 27*(10), 1032–43.

Hirsch, R.L., Bezdek, R., and Wendling, R. (2005). *Peaking of World Oil Production. Impacts, Mitigation, and Risk Management.* Pittsburgh, PA: U.S. Department of Energy National Energy Technology Laboratory.

Holdway, A.R., Williams, A.R., Inderwildi, O.R., and King, D.A. (2010). Indirect emissions from electric vehicles: Emissions from electricity generation. *Energy and Environmental Science, 3,* 1825–32.

Horst, J., Frey, G., and Leprich, U. (2009). *Auswirkungen von Elektroautos auf den Kraftwerkspark und die CO2-Emissionen in Deutschland* [Effects of electric cars on the power plant fleet and CO2-emissions in Germany]. Frankfurt am Main, Germany: WWF Deutschland.

Huo, H., Zhang, Q., Wang, M.Q., Streets, D.G., and He, K. (2010). Environmental implication of electric vehicles in China. *Environmental Science and Technology, 44,* 4856–61.

Karrer, K., Glaser, C., Clemens, C., and Bruder, C. (2009). Technikaffinität erfassen - der Fragebogen TA-EG [Measure affinity to technology: The questionnaire TA-EG]. In A. Lichtenstein, C. Stößel and C. Clemens (Eds.), *Der Mensch im Mittelpunkt technischer Systeme 8. Berliner Werkstatt Mensch-Maschine-Systeme 7. bis 9. Oktober 2009 (ZMMS* Spektrum *Band 22)* (vol. 29, pp. 196–201) (Düsseldorf: VDI).

Krems, J.F., Franke, T., Neumann, I., and Cocron, P. (2010). Research methods to assess the acceptance of EVs: Experiences from an EV user study. In T. Gessner (Ed.), *Smart Systems Integration: 4th European Conference and Exhibition on Integration Issues of Miniaturized Systems: MEMS, MOEMS, ICs and Electronic Components. Como, Italy.* Berlin, Germany: VDE.

Kunert, U., and Follmer, R. (2005). Methodological advances in national travel surveys: Mobility in Germany 2002. *Transport Reviews, 25,* 415–31.

Kurani, K.S., Turrentine, T., and Sperling, D. (1994). Demand for electric vehicles in hybrid households: An exploratory analysis. *Transport Policy, 1*, 244–56.

Kurani, K.S., Turrentine, T., and Sperling, D. (1996). Testing electric vehicle demand in 'hybrid households' using a reflexive survey. *Transportation Research Part D: Transport and Environment, 1*, 131–50.

Nemeth, C. (2003). *Human factors methods for design: Making systems human-centered.* London: Taylor and Francis.

Neumann, I., Cocron, P., Franke, T., and Krems, J.F. (2010). Electric vehicles as a solution for green driving in the future? A field study examining the user acceptance of electric vehicles. Proceedings of the *European Conference on Human Interface Design for Intelligent Transport Systems, Berlin, Germany.*

Rögele, S., Schweizer-Ries, P., Zoellner, J., and Antoni, C.H. (2010). Abschlussbericht MINI.E. Berlin powered by Vattenfall - Betrachtung der Aspekte Umweltbewusstsein, erneuerbare Energien, öffentliche Ladesäulen und gesteuertes Laden bei Elektromobilität [Final report MINI E Berlin powered by Vattenfall: Consideration of the aspects environmental awareness, renewable energies, public charging stations and managed charging in electric mobility]. Trier: University of Trier.

Schafer, A. (1998). The global demand for motorized mobility. *Transportation Research Part A: Policy and Practice, 32*, 455–77.

Thiel, C., Perujo, A., and Mercier, A. (2010). Cost and CO_2 aspects of future vehicle options in Europe under new energy policy scenarios. *Energy Policy, 38*, 7142–51.

Van Der Laan, J.D., Heino, A., and De Waard, D. (1997). A simple procedure for the assessment of acceptance of advanced transport telematics. *Transportation Research Part C: Emerging Technologies, 5*, 1–10.

UN General Assembly. *Resolution adopted by the General Assembly: 60/1 – 2005 World Summit Outcome.* [Online]. Available at: http://daccess-ods.un.org/TMP/959501.2.html, accessed 31 January 2011.

Index

Bold page numbers indicate figures, *italic* numbers indicate tables